GD&T

Application and Interpretation

Sixth Edition

by

Bruce A. Wilson

Publisher

The Goodheart-Willcox Company, Inc.

Tinley Park, IL

www.g-w.com

The Goodheart-Willcox Company, Inc. Brand Disclaimer: Brand names, company names, and illustrations for products and services included in this text are provided for educational purposes only and do not represent or imply endorsement or recommendation by the author or the publisher.

The Goodheart-Willcox Company, Inc. Safety Notice: The reader is expressly advised to carefully read, understand, and apply all safety precautions and warnings described in this book or that might also be indicated in undertaking the activities and exercises described herein to minimize risk of personal injury or injury to others. Common sense and good judgment should also be exercised and applied to help avoid all potential hazards. The reader should always refer to the appropriate manufacturer's technical information, directions, and recommendations; then proceed with care to follow specific equipment operating instructions. The reader should understand these notices and cautions are not exhaustive.

The publisher makes no warranty or representation whatsoever, either expressed or implied, including but not limited to equipment, procedures, and applications described or referred to herein, their quality, performance, merchantability, or fitness for a particular purpose. The publisher assumes no responsibility for any changes, errors, or omissions in this book. The publisher specifically disclaims any liability whatsoever, including any direct, indirect, incidental, consequential, special, or exemplary damages resulting, in whole or in part, from the reader's use or reliance upon the information, instructions, procedures, warnings, cautions, applications, or other matter contained in this book. The publisher assumes no responsibility for the activities of the reader.

The Goodheart-Willcox Company, Inc. Internet Disclaimer: The Internet resources and listings in this Goodheart-Willcox Publisher product are provided solely as a convenience to you. These resources and listings were reviewed at the time of publication to provide you with accurate, safe, and appropriate information. Goodheart-Willcox Publisher has no control over the referenced websites and, due to the dynamic nature of the Internet, is not responsible or liable for the content, products, or performance of links to other websites or resources. Goodheart-Willcox Publisher makes no representation, either expressed or implied, regarding the content of these websites, and such references do not constitute an endorsement or recommendation of the information or content presented. It is your responsibility to take all protective measures to guard against inappropriate content, viruses, or other destructive elements.

Introduction

This study guide has been written to supplement the *GD&T: Application and Interpretation* textbook. The review questions and application problems contained in this study guide can be completed on the basis of the information provided by the textbook. Other textbooks may be used, but it is unlikely that any other textbook will provide all the information necessary to answer all the questions or work all the application problems.

The textbook and this study guide used together to provide the information and practice necessary to gain a strong working knowledge of dimensioning and tolerancing practices.

A majority of the material in the textbook and the study guide requires an understanding of only basic mathematics. Some of the material requires simple algebra operations, such as solving for one unknown value when two known values are provided. Knowledge of print reading or basic drafting techniques will be helpful in understanding the illustrations and completing application problems.

To get the maximum benefit from the textbook and study guide materials, the following study methods are recommended.

1. Read the objectives at the beginning of each chapter of the study guide prior to reading the corresponding chapter in the textbook.
2. As you read the textbook chapter, make a list of questions regarding information that is not understood.
3. Complete the review questions and application problems after reading the textbook material.
4. Cross off the questions from step 2 and 3 as answers are provided during a classroom presentation. Ask the instructor to provide answers if the presentation does not provide all the answers to your questions.
5. Correct the answers to your review questions and application problems on the basis of classroom reviews. The corrected materials will be useful for studying for exams.

The objectives at the beginning of each chapter in this study guide define what you should be able to do after studying the textbook, completing outside study activities, attending classroom lectures, and completing study guide review questions and application problems. The level of achievement will depend to a great extent on the amount of time devoted to studying the textbook and study guide materials. Full mastery of dimensioning and tolerancing methods requires studying the fundamentals, then applying them to real industrial applications.

Individuals who put forth the effort to become proficient in dimensioning and tolerancing methods and use that ability to maximize clarity of product design requirements and provide maximum permissible tolerances will be rewarded with the satisfaction of knowing that they are producing the best possible results.

Bruce A. Wilson

Table of Contents

Chapter 1

Introduction to Dimensioning and Tolerancing

Name _____ Date _____ Class _____

Reading

Read Chapter 1 of the *GD&T: Application and Interpretation* textbook prior to completing the review exercises.

Objectives

A combination of activities is required to achieve the following objectives. Completing the reading assignment and the following review exercises is an important part of achieving the objectives. Familiarization with the objectives prior to completion of the reading assignment and review exercises will make mastery of the objectives easier. After completing the reading assignment and completing the review exercises, you will be able to:

▼ Explain the importance of accurately specifying dimensions and tolerances.
▼ Recall the history and development of dimensioning and tolerancing methods.
▼ Explain how teamwork can result in better definition of the dimensions and tolerances shown on a drawing or in a computer-aided design (CAD) file.
▼ Recall the job titles of those who should be on the design process team.
▼ Recall the dimensioning and tolerancing skills needed for success in design- or production-related occupations.
▼ Analyze some possible industrial changes and the impacts of these changes on dimensioning and tolerancing.
▼ Understand how views are created using orthographic projection.

Review Exercises

Place your answers in the spaces provided. Show all calculations for problems that require mathematical solutions.

Multiple Choice

_____ 1. The wavelength of a specific color of light is used in determining the length of one _____.
A. foot
B. yard
C. meter
D. kilometer

_____   2. A(n) _____ is responsible for dimensioning a part in such a way that the functional needs are met and the part is producible.
    A. designer
    B. inspector
    C. production planner
    D. machinist

_____   3. Tolerance values should be _____.
    A. assigned to meet the desires of manufacturing
    B. assigned on the basis of what worked on prior designs
    C. selected from a table in ASME Y14.5
    D. calculated to ensure proper function of the design with consideration given to manufacturing capabilities

_____   4. The _____ system may be used for accurate measurements.
    A. metric
    B. inch
    C. Both A and B.

_____   5. The preferred metric value for dimensions on a mechanical drawing is _____.
    A. millimeters
    B. centimeters
    C. meters
    D. kilometers

_____   6. A machinist might be able to help a designer by telling him or her _____.
    A. the size tool needed to produce a particular feature
    B. the tolerance that is achievable
    C. about machine capability
    D. All of the above.

_____   7. One method of reducing the number of unnecessarily small tolerances is to _____ tolerances.
    A. double the value of all assumed
    B. calculate all
    C. remove
    D. None of the above.

_____   8. Part requirements _____ if dimensions and tolerances are in compliance with the standard.
    A. are confusing
    B. are poorly defined
    C. become difficult to meet
    D. have well-defined meanings

_____   9. The application of _____ on a drawing defines the amount of acceptable variation on a dimensioned feature.
    A. dimensions
    B. notes
    C. tolerances
    D. None of the above.

Name _____

True/False

_____ 10. *True or False?* The current standard specifies that all measurements must be in inches.

_____ 11. *True or False?* The designer should work independent of others to achieve an optimum design.

_____ 12. *True or False?* The symbol for inches must be applied to all values less than one inch.

_____ 13. *True or False?* Disagreement about drawing requirements can occur when nonstandard dimensioning methods are used.

_____ 14. *True or False?* Interpretation of a drawing is the ability to determine part requirements from what is shown when the drawing complies with standards.

_____ 15. *True or False?* A projection symbol should be included on orthographic drawings to indicate whether the views are created using first or third angle projection.

Fill in the Blank

_____ 16. A(n) _____ is an ancient unit of measurement based on the distance across a finger.

_____ 17. _____ may be used to establish or show relationships between features in adjacent orthographic views.

Short Answer

18. Why is it important to have an accurate distance standard?

_____

_____

19. Give one reason why nonstandard symbols are generally avoided.

_____

_____

20. Show a note that should be placed on a drawing that primarily has inch dimensions.

_____

_____

21. Why is it important for an inspector to correctly interpret the dimensions on a drawing?

_____

_____

22. When is it necessary to know the requirements of a previous issue of the ASME Y14.5 dimensioning and tolerancing standard?

_____

_____

23. What has made it possible for all paper drawings to be eliminated from a factory?

_____

_____

Chapter 2

Dimensioning and Tolerancing Symbology

Name _____ Date _____ Class _____

Reading

Read Chapter 2 of the *GD&T: Application and Interpretation* textbook prior to completing the review exercises.

Objectives

A combination of activities is required to achieve the following objectives. Completing the reading assignment and the following review exercises is an important part of achieving the objectives. Familiarization with the objectives prior to completion of the reading assignment and review exercises will make mastery of the objectives easier. After completing the reading assignment and completing the review exercises, you will be able to:

▼ Identify and draw general dimensioning symbols and show their general applications.
▼ Identify and draw tolerancing symbols and show their general applications.
▼ Complete a feature control frame using the correct order of segments in the frame.
▼ Identify basic dimensions and define means for indicating a basic dimension on a drawing or in a design model.

Review Exercises

Place your answers in the spaces provided. Accurately complete any required sketches. Show all calculations for problems that require mathematical solutions.

Multiple Choice

_____ 1. A value shown _____ is a reference value.
   A. in brackets
   B. underlined
   C. with an arc above it
   D. in parentheses

_____ 2. The origin symbol is _____.
   A. applied to one end of all dimensions
   B. applied to both ends of some dimensions
   C. applied in place of one arrowhead
   D. never used

_____   3. In CAD models and on drawings, _____ are being replaced by symbols.
A. abbreviations
B. nonstandard symbols
C. notes
D. None of the above.

_____   4. Symbols in a CAD system are generally _____ to save time when dimensioning.
A. made part of a font or library of symbols
B. drawn to approximate dimensions
C. omitted
D. None of the above.

_____   5. Present practice requires the radius symbol be _____ the dimension value.
A. placed after
B. placed in front of
C. larger than the characters in
D. smaller than the characters in

_____   6. The _____ tolerances are straightness, circularity, flatness, and cylindricity.
A. position
B. orientation
C. form
D. runout

_____   7. Feature control frames _____.
A. have a required format
B. may be formatted by personal preference
C. vary between companies
D. None of the above.

_____   8. Angularity is a type of _____ tolerance.
A. form
B. orientation
C. position
D. profile

True/False

_____   9. _True or False?_ The preferred method to show a depth specification is to use an abbreviation for depth.

_____   10. _True or False?_ Ambiguous tolerance specifications can be the result of using nonstandard symbols.

_____   11. _True or False?_ The abbreviation CBORE and the symbol for counterbore may be used on the same drawing.

_____   12. _True or False?_ The current standard requires that datum feature references in a feature control frame be located between the tolerance symbol and the tolerance value.

Name _____

_____ 13. *True or False?* A diameter symbol is placed in front of the tolerance value in all feature control frames.

_____ 14. *True or False?* A datum feature symbol, in an orthographic view, may be applied on either side of an extension line without affecting the meaning of the symbol.

_____ 15. *True or False?* Symbols are required to be sized proportional to the feature to which the symbols are applied.

_____ 16. *True or False?* Tolerance symbols are generally shaped to give an indication of the required control.

_____ 17. *True or False?* Abbreviations and words are typically used in notes lists, but symbols may be used in notes.

_____ 18. *True or False?* All feature control frames must show a material condition modifier following the tolerance value.

Fill in the Blank

_____ 19. Using symbols _____ the number of words that are placed on a drawing.

_____ 20. There are _____ different form tolerance symbols.

_____ 21. Feature control frames specify tolerances to be applied to _____ or features of size.

_____ 22. Any tolerance applied to a thread and shown in a feature control frame is assumed to apply to the _____ diameter of the thread unless indicated otherwise.

_____ 23. A(n) _____ may be used to indicate that all dimensions are basic.

_____ 24. A(n) _____ dimension can be indicated by drawing a rectangle around the dimension value.

_____ 25. The abbreviation for *regardless of feature size* is _____.

Short Answer

26. The letter X may be used as a symbol. What are the two possible uses of the symbol X?

_____

_____

27. Explain how each of the meanings for the symbol X is indicated.

_____

_____

28. How is the symbol size determined for a drawing?

_____

_____

29. If a drawing is being produced by hand, what is one method of ensuring that symbols are quickly drawn and correctly sized?

_____

_____

30. List the two types of profile tolerance symbols. The names of the symbols must be given.

_____

_____

31. Describe the total runout symbol.

_____

_____

32. What is the order in which datums are referenced?

_____

_____

33. List the three datum target types.

_____

_____

Name _____

Application Problems

All application problems are to be completed using correct dimensioning techniques. Show any required calculations.

34. Show the diameter symbol in the correct location on each of the diameter dimensions.

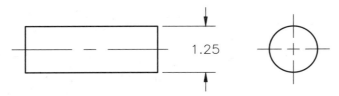

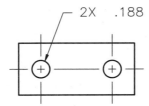

35. Properly show the radius symbol on each of the radius dimensions.

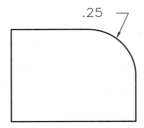

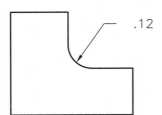

36. Show the spherical diameter symbol on the given dimension.

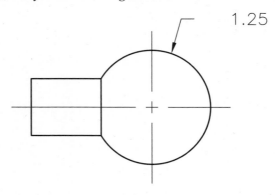

1.25

37. Use symbols to complete the hole and counterbore specification. Correctly position the specification near the leader.

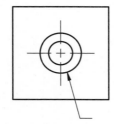

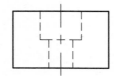

.250 DIA HOLE
.431 DIA COUNTERBORE
.31 DEEP
.03 CORNER RADIUS

38. Use symbols to complete the hole and countersink specification. Correctly position the specification near the leader.

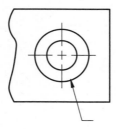

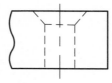

2 HOLES
.313 DIAMETER HOLES
.562 DIA COUNTERSINK
82° COUNTERSINK

Name _____

39. Label each compartment of the feature control frame.

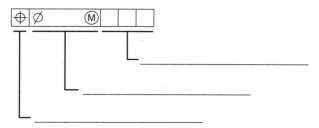

_____

_____

_____

40. Identify each of the given symbols.

A. ⌀ _____

B. ⌴ _____

C. ∨ _____

D. ⟱ _____

E. A◀ _____

F. ⊕ _____

G. — _____

H. ⁄⁊ _____

I. ⊥ _____

J. ∠ _____

K. // _____

L. ⌒ _____

M. ⬈ _____

N. ⬈⬈ _____

O. Ⓜ _____

P. Ⓛ _____

Q. ⊖ _____

NOTES

Chapter 3
General Dimensioning Requirements

Name _____ Date _____ Class _____

Reading

Read Chapter 3 of the *GD&T: Application and Interpretation* textbook prior to completing the review exercises.

Objectives

A combination of activities is required to achieve the following objectives. Completing the reading assignment and the following review exercises is an important part of achieving the objectives. Familiarization with the objectives prior to completion of the reading assignment and review exercises will make mastery of the objectives easier. After completing the reading assignment and completing the review exercises, you will be able to:

▼ Apply general dimensioning methods using the correct line types, lettering sizes, and arrowhead form.
▼ Describe and apply general dimensioning systems including chain, baseline, rectangular coordinate, and polar coordinate dimensions.
▼ Utilize preferred dimension placement to provide clear part requirements specification.
▼ Apply general and specific notes on a drawing.
▼ Cite the general categories of fit between mating parts.

Review Exercises

Place your answers in the spaces provided. Show all calculations for problems that require mathematical solutions.

Multiple Choice

_____ 1. Extension lines begin approximately ____ from the dimensioned feature to provide a visible gap.
A. .015″
B. .062″
C. .125″
D. .188″

2. Extension lines extend approximately _____ past the last dimension line.
    A. .020"
    B. .062"
    C. .125"
    D. .188"

3. Extension lines may be broken where they cross _____.
    A. other extension lines
    B. object lines
    C. hidden lines
    D. arrowheads

4. The recommended minimum distance between adjacent dimensions is _____.
    A. .12"
    B. .24"
    C. .31"
    D. .44"

5. _____ dimensions have all values written horizontally.
    A. Aligned
    B. Unidirectional
    C. Metric
    D. Inch

6. A zero is placed in front of values less than 1.00 when using _____.
    A. aligned dimensions
    B. unidirectional dimensions
    C. metric values
    D. inch values

7. Tolerance _____ can be affected by whether chain or baseline dimensions are applied to a part.
    A. interpretation
    B. accumulation
    C. values
    D. None of the above.

8. Tabulated dimensions can be used to specify _____.
    A. location
    B. size
    C. tolerances
    D. All of the above.

9. Typically, a dimension line is perpendicular to a(n) _____ line.
    A. object
    B. extension
    C. hidden
    D. leader

Name _____

_____ 10. Adjacent dimension values are normally _____ to make them easier
to read.
   A. offset
   B. lined up
   C. avoided
   D. None of the above.

_____ 11. A(n) _____ view sometimes requires that one end of a dimension apply to
a hidden feature.
   A. profile
   B. auxiliary
   C. full section
   D. half section

_____ 12. _____ dimensioning is applying dimensions in such a manner as to
result in more than one means of defining the dimension and tolerance
on a feature.
   A. Double
   B. Duplicate
   C. Ordinate
   D. Third angle

_____ 13. A dimension value placed _____ indicates the value is for reference only.
   A. between quotation marks
   B. inside a rectangle
   C. between parentheses
   D. between brackets

_____ 14. When the maximum shaft size is equal to the minimum hole size, the
mating parts have a zero _____.
   A. transition
   B. material condition
   C. tolerance
   D. allowance

True/False

_____ 15. *True or False?* Size dimensions define the location of features.

_____ 16. *True or False?* The unidirectional dimensioning system usually requires
more space for vertical dimensions than does the aligned dimensioning
system.

_____ 17. *True or False?* Regardless of the drawing scale, the drawing must show
the dimension values to be produced.

_____ 18. *True or False?* Visualizing the geometric shapes in a part can help
determine what dimensions are needed.

_____ 19. *True or False?* The view in which a feature is dimensioned may be selected
at random.

_____ 20. *True or False?* Dimensioning between views is a good practice that makes it easier to relate dimensions to two views.

_____ 21. *True or False?* Dimensions to hidden features are common since many holes are shown with hidden lines.

_____ 22. *True or False?* When possible, all dimensions should be placed on a view in which the dimensioned features are seen in true size.

_____ 23. *True or False?* General notes provide information that applies to the entire drawing.

_____ 24. *True or False?* Notes must be shown on the drawing sheets that contain the views of the part.

Fill in the Blank

_____ 25. A leader line has an arrowhead on _____ end.

_____ 26. The recommended minimum distance from an object to the first dimension line is _____.

_____ 27. Notes are connected to features using a _____.

_____ 28. The recommended length-to-width ratio for an arrowhead is _____.

_____ 29. The _____ dimensioning system has values aligned with the dimension lines.

_____ 30. Rectangular coordinate dimensioning without dimension lines places dimension values at the ends of _____ lines.

_____ 31. Polar dimensions include a distance and _____.

_____ 32. A(n) _____ used to replace one of the arrowheads on a dimension line indicates the origin for the dimension.

_____ 33. A(n) _____ dimension can be indicated by drawing a rectangle around the number.

_____ 34. A feature _____ frame contains a tolerance that is attached to a feature.

Short Answer

35. When may a leader line be broken?

_____

_____

Name _____

36. List two of the possible arrangements for arrowheads and dimension values in relationship to the extension lines.

_____

_____

_____

37. Why are horizontal and vertical leader lines avoided?

_____

_____

38. Describe an advantage of using unidirectional dimensioning over aligned dimensioning in orthographic views.

_____

_____

39. When is it necessary to show the unit of measurement for a dimension?

_____

_____

40. Why are larger dimensions typically placed outside smaller dimensions?

_____

_____

_____

41. Where may section lines be broken to make dimension application in a section view more clear?

_____

_____

Application Problems

All application problems are to be completed using correct dimensioning techniques. Show any required calculations.

42. Show the symbol for each of the following:

    A. Maximum material condition _____

    B. Least material condition _____

43. Circle the dimension value for each of the *size* dimensions.

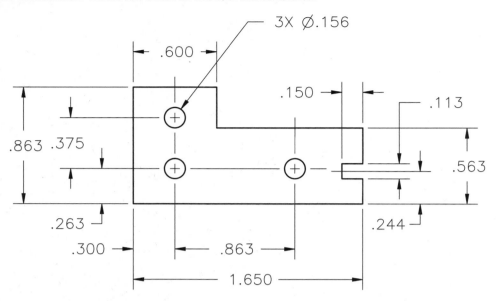

44. In place of each of the question marks, indicate the recommended value for dimensioning.

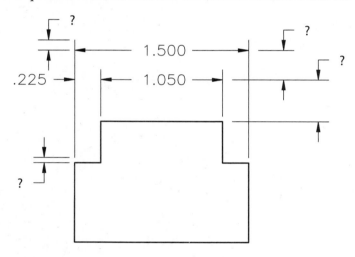

45. Apply dimension values to the shown slot using unidirectional dimensions. The slot is .250″ wide and .125″ deep.

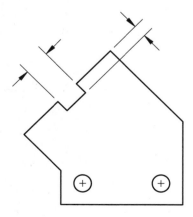

Name _____

46. A full scale and half scale drawing of the same rectangular part are given. Dimension both of the drawings. The actual size of the rectangle is 2.00″ × 1.00″.

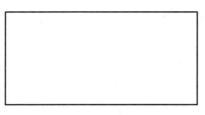

SCALE: 1/1

SCALE: 1/2

47. What are the maximum and minimum permissible horizontal dimensions between points A and F on a part produced to the given drawing? Assume no form or orientation variation exists to complicate the problem.

_____ Maximum

_____ Minimum

TOLERANCES:
.XX = ±.02
.XXX = ±.005

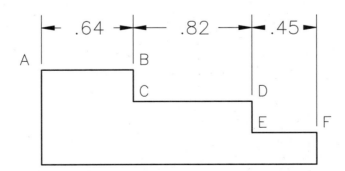

48. What are the maximum and minimum permissible horizontal dimensions between points C and D on a part produced to the given drawing? Assume no form or orientation variation exists to complicate the problem.

_____ Maximum

_____ Minimum

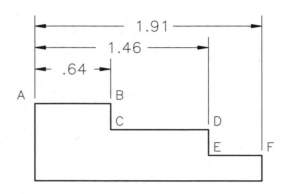

TOLERANCES:
.XX = ±.02
.XXX = ±.005

49. What is the specified size for hole B1 and what is the allowable size variation?

_____ Specified size

_____ Allowable size variation

What is the coordinate location for hole B1?

_____ X Coordinate location

_____ Y Coordinate location

What is the specified size for hole A2 and what is the allowable size variation?

_____ Specified size

_____ Allowable size variation

What is the coordinate location for hole A2?

_____ X Coordinate location

_____ Y Coordinate location

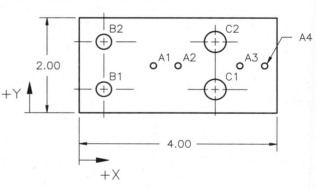

DRILL TABLE

SYMBOL	LOCATION		SIZE	TOL
	+X	+Y		
A1	1.50	1.00		
A2	2.00	1.00	.125	+.005
A3	3.25	1.00		−.000
A4	3.75	1.00		
B1	.50	.50	.312	+.005
B2	1.00	1.50		−.000
C1	2.75	.50	.438	+.006
C2	2.75	1.50		−.000

Name _____

50. Locate vertex A for the inclined surface and dimension the angle.

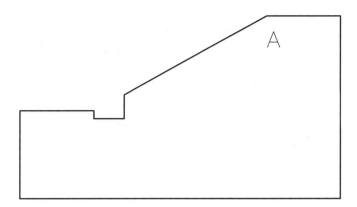

51. Apply dimensions to the given part. Be certain to apply dimensions where the feature profiles are best shown. Use appropriate spacing between dimensions.

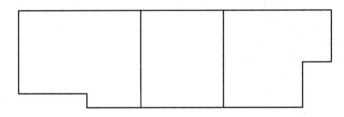

52. Dimension all features.

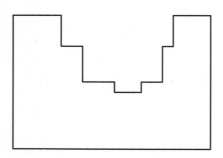

53. Dimension the depth for each slot. Also dimension the location of the hole.

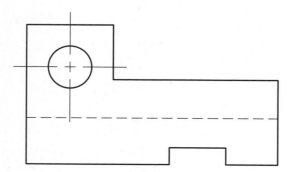 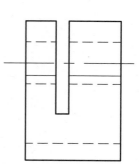

Name _____

54. Dimension the given section view and add section lining (crosshatching).

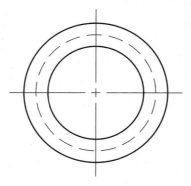

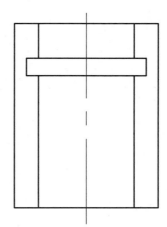

55. Apply 1.0003″ and 1.0000″ limits of size to the outside diameter.

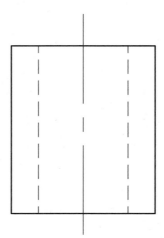

56. Apply the needed dimensions on the orthographic views.

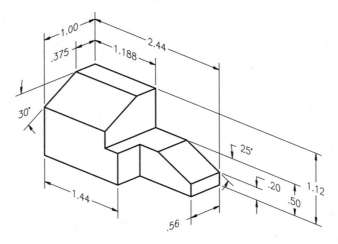

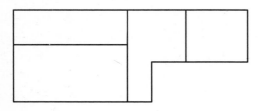

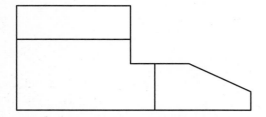

Name _____

57. Apply the needed dimensions on the orthographic views.

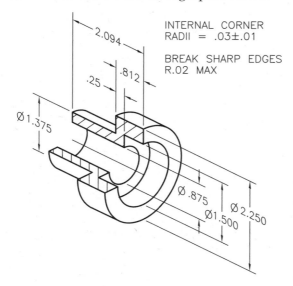

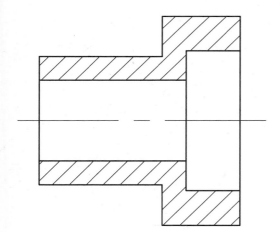

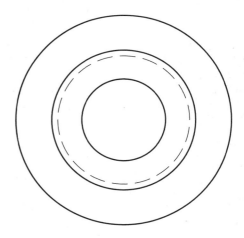

58. Apply the needed dimensions on the views provided.

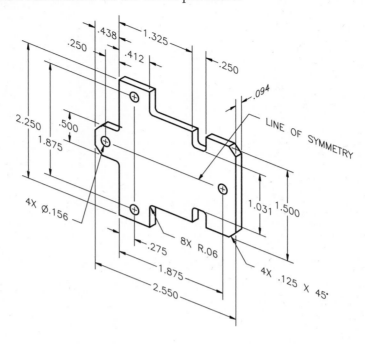

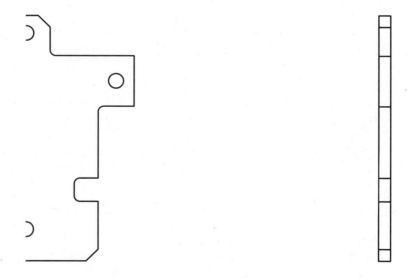

Chapter 4

Dimension Application and Limits of Size

Name _____   Date _____   Class _____

Reading

Read Chapter 4 of the *GD&T: Application and Interpretation* textbook prior to completing the review exercises.

Objectives

A combination of activities is required to achieve the following objectives. Completing the reading assignment and the following review exercises is an important part of achieving the objectives. Familiarization with the objectives prior to completion of the reading assignment and review exercises will make mastery of the objectives easier. After completing the reading assignment and completing the review exercises, you will be able to:

▼ Clearly apply dimensions by complying with the stated general dimensioning guidelines.
▼ Apply dimensions to any of the geometric shapes commonly found on mechanical parts.
▼ Cite the categories for limits of fit and describe the general condition created by each category.
▼ Calculate and apply limits of size for mating features.
▼ Explain Rule #1 and Rule #2 of the ASME Y14.5-2009 standard.
▼ Provide examples of the effects that dimensions and tolerances have on manufacturing.
▼ Complete a surface texture specification when provided the allowable variations.

Review Exercises

Place your answers in the spaces provided. Show all calculations for problems that require mathematical solutions.

Multiple Choice

_____   1. An angle in an orthographic view is assumed to be _____ when lines are drawn perpendicular to one another.
    A. untoleranced
    B. basic
    C. 90°
    D. No assumption permitted.

2. A right circular cone is dimensioned by giving the base diameter and
    _____.
    A. cone height
    B. cone angle
    C. Either A or B.
    D. Neither A nor B.

3. Leaders extending from a hole specification should point toward the
    _____ of the hole when connected to the circular view of the hole.
    A. center
    B. vertical centerline
    C. horizontal centerline
    D. Either B or C.

4. Hole locations are dimensioned to the _____ of the hole.
    A. edge
    B. bottom
    C. end
    D. center

5. If two groups of holes have sizes that are close to the same diameter, all
    holes of one diameter may be _____ to make it possible to tell the size of
    all holes.
    A. labeled
    B. drawn out of scale
    C. omitted
    D. None of the above.

6. Specify a counterbore hole by giving the hole diameter (when required),
    hole depth, and _____.
    A. corner radius
    B. counterbore depth
    C. counterbore diameter
    D. All of the above.

7. A common use for a _____ is to provide a recess for a flathead screw.
    A. counterbore
    B. countersink
    C. counterdrill
    D. None of the above.

8. A spotface depth may be specified by _____.
    A. noting the depth
    B. dimensioning the remaining material
    C. Either A or B.
    D. Neither A nor B.

9. Angles are typically dimensioned using values expressed in _____.
    A. degrees
    B. radians
    C. arc lengths
    D. None of the above.

Name _____

_____ 10. The letter R in a radius dimension is shown as a _____ to the dimension value.
A. prefix
B. suffix
C. Either A or B.
D. Neither A nor B.

_____ 11. Although no break is required, extension lines may be broken where they cross _____.
A. extension lines
B. dimension lines
C. object lines
D. arrowheads

_____ 12. The minimum allowable bend radius for a sheet metal part is affected by the _____.
A. type of material
B. hardness condition of the material
C. material thickness
D. All of the above.

_____ 13. A bend radius that is too small can result in _____ that weaken the part.
A. ridges
B. sharp edges
C. cracks
D. None of the above.

_____ 14. The maximum limit of size is placed _____ the minimum limit of size when shown in a dimension.
A. below
B. above
C. to the right of
D. to the left of

_____ 15. When using the _____ system, the limits of size for the shaft are calculated to fit the hole.
A. basic tolerancing
B. position tolerancing
C. basic hole
D. basic shaft

_____ 16. A clearance fit used for moving parts is designated by the letters _____.
A. RC
B. LC
C. LT
D. FN

_____ 17. Which of the following classes of fit is most likely to result in a clearance condition?
A. LT1
B. LT6
C. LN2
D. FN4

_____ 18. Which rule in ASME Y14.5 requires perfect form at MMC?
  A. Rule #1
  B. Rule #2
  C. Rule #3
  D. Rule #4

_____ 19. _____ specifications define allowable variations known as roughness, waviness, and lay.
  A. Limit of size
  B. Surface conditions
  C. Form tolerance
  D. Class of fit

_____ 20. The distance across peaks and valleys that cause surface roughness is known as the _____.
  A. waviness
  B. roughness width
  C. roughness distance
  D. surface texture

True/False

_____ 21. _True or False?_ Dimensions to completely define a pyramid are the base dimensions and the apex location dimensions.

_____ 22. _True or False?_ Holes are normally dimensioned by giving the radius.

_____ 23. _True or False?_ A large hole may be dimensioned with the dimension line, arrowheads, and dimension value located within the circle that represents the hole.

_____ 24. _True or False?_ The depth specification for a hole is the distance to the end of the drill point.

_____ 25. _True or False?_ Hole depth should be shown in front of the hole diameter in a hole size specification.

_____ 26. _True or False?_ The dimension line for an angle is drawn as an arc with the center located at the vertex of the angle formed by the extension lines.

_____ 27. _True or False?_ Arcs should be dimensioned in a view where they are foreshortened rather than in a true shape view.

_____ 28. _True or False?_ A centerdrilled hole in the end of a shaft, when used in a machine setup, locates the center (or axis) of the shaft.

_____ 29. _True or False?_ Every feature of size has a minimum and maximum allowable size, even when a single limit dimension is applied to the feature.

_____ 30. _True or False?_ An RC1 class of fit results in less clearance than an RC4 class of fit.

Name _____

_____ 31. *True or False?* Fabrication accuracy capabilities and methods do not generally need to be considered when applying dimensions or calculating tolerances.

_____ 32. *True or False?* The lifecycle costs for mated assemblies can be higher than for interchangeable assemblies.

Fill in the Blank

_____ 33. The diameter and _____ dimension must be given for a cylindrical part.

_____ 34. A diameter dimension line applied on a circular view is oriented to pass through the _____ of the dimensioned feature.

_____ 35. The abbreviation for counterbore is _____.

_____ 36. A countersink hole specification includes a hole diameter, countersink _____, and countersink angle.

_____ 37. What is the equivalent decimal degree value for 25°30′?

_____ 38. Chamfers made at a(n) _____ angle may be dimensioned with a note.

_____ 39. The leader for a radius dimension extends through the arc _____.

_____ 40. Limit dimensions specify the _____ and _____ acceptable dimension values.

_____ 41. When using the basic _____ system for calculation of size limits, one size limit for the shaft is the basic size.

_____ 42. _____ is the direction of tool marks, scratches, or the grain may be specified as part of a surface texture specification.

Short Answer

43. Explain how a single view can be dimensioned to completely define a cylindrical part.

_____

_____

_____

_____

44. What is the effect of using very small size tolerances?

45. If a pattern of holes is repeated several times on a drawing, why would a removed view be used to define the hole locations within the pattern?

46. Define counterbore and list one application of a counterbore.

47. How deep must a spotface be made if no depth dimension is shown?

48. How is a centerline identified as a line of symmetry?

49. What are four pieces of information that must be included in a thread specification?

50. How can an exception to Rule #1 be specified?

Name _____

Application Problems

All application problems are to be completed using correct dimensioning technique. Show any required calculations.

51. On the orthographic view provided below, apply all dimensions necessary to completely define the given part.

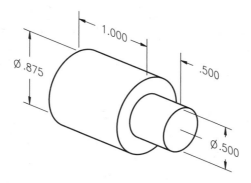

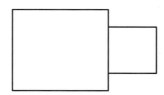

52. Apply diameter dimensions to the given holes.

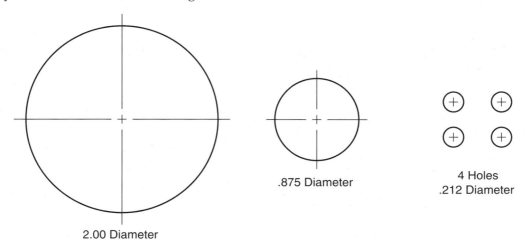

2.00 Diameter

.875 Diameter

4 Holes
.212 Diameter

53. Apply a hole specification to the given hole using the given information. Use symbology.

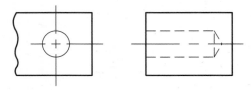

Hole diameter: .188 +.006 − .003
Depth: .500±.010

54. Dimension each of the following angles.

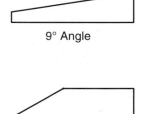

9° Angle

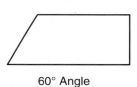

30° Angle

60° Angle

55. Completely dimension each part, estimating dimension values. The arc on one of the parts must be located by dimensioning the tangents. The arc on the other part must be located by dimensioning the arc center. Do not double dimension any feature.

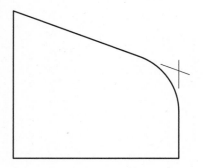

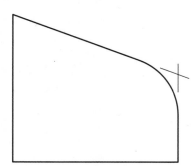

Name _____

56. Dimension each slot using the dimension values provided.

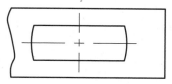

Length: 1.00
Width: .38
Radius: .75
Corner Radius: .03 MAX

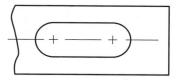

Length: 1.00
Width: .38
Radius: .19

57. Dimension the shaft diameter and the keyseat. Use the dimension information provided.

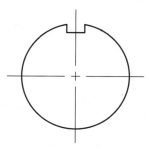

Shaft diameter: 1.125
Keyseat width: .1875
Keyseat depth: .0938

58. Completely dimension the sheet metal part. Estimate dimension values.

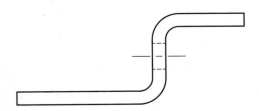 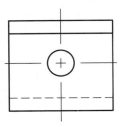

59. Dimension the slot width on each of the given drawings. Use limit dimensions on the indicated part and plus or minus tolerances on the other part. Determine dimension values from the shown information.

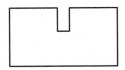

Limit dimension
Slot width: .125
Plus tolerance: .005
Minus tolerance: .002

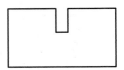

Plus or minus tolerance
Slot width: .125
Plus tolerance: .005
Minus tolerance: .002

Name _____

60. Add the necessary information to the drawing to permit exception to the requirements of Rule #1 for the thickness dimension.

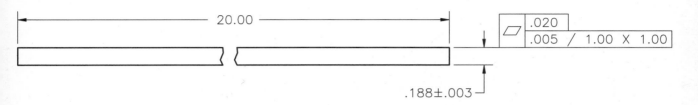

61. Complete the given drawing by entering the information for a revision. The indicated hole was previously dimensioned as a .250″ diameter. It is now to be a .261″ diameter with a .006″ plus tolerance and .003″ minus tolerance. Also complete the revision block.

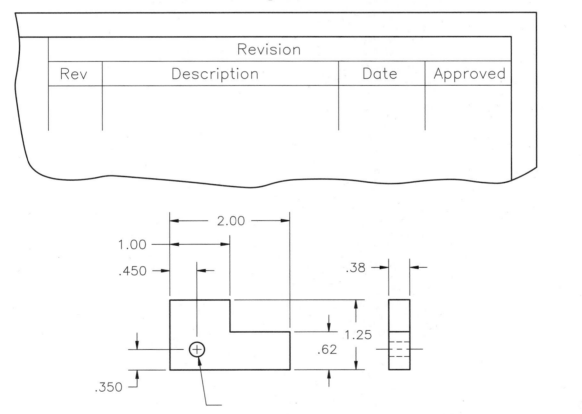

62. Calculate limits of size and apply dimensions for the shown parts. Show all calculations. (See Figure 4-47 of the textbook.) Use tolerance tables in ASME B4.1 or Machinery's Handbook. Apply the dimensions using limit dimensions.

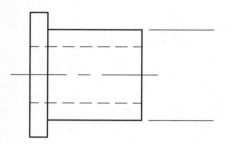
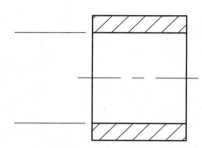

Basic hole system
Basic size: 1.000
Class of fit: LC3

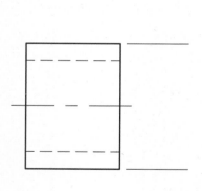
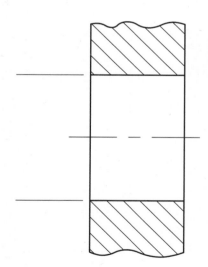

Basic hole system
Basic size: 1.375
Class of fit: FN4

Name _____

63. Calculate limits of size for the shaft and hole. Show all calculations. Split the allowable tolerance evenly between the two parts.

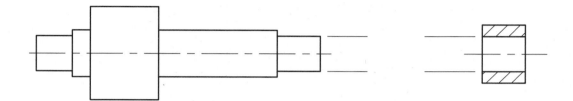

Basic hole system
Basic size: .375
Allowance: .0004
Maximum clearance: .0022

64. Complete a surface texture specification that permits a roughness of 125 microinches with a roughness width cutoff value of .03". No lay direction is required.

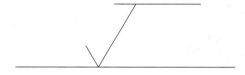

65. Complete a surface texture specification that permits a minimum roughness of 63 microinches and a maximum roughness of 250 microinches with a roughness width cutoff of .100". No additional control is needed.

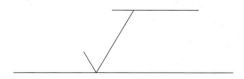

NOTES

Chapter 5

Form Tolerances

Name _____ Date _____ Class _____

Reading

Read Chapter 5 of the *GD&T: Application and Interpretation* textbook prior to completing the review exercises.

Objectives

A combination of activities is required to achieve the following objectives. Completing the reading assignment and the following review exercises is an important part of achieving the objectives. Familiarization with the objectives prior to completion of the reading assignment and review exercises will make mastery of the objectives easier. After completing the reading assignment and completing the review exercises, you will be able to:

▼ Draw the symbols for form tolerances.
▼ Complete a feature control frame to specify a form tolerance and properly apply material condition modifiers on the tolerances.
▼ Explain the extent of form control established by limits of size.
▼ Apply straightness tolerances to control surface elements or a derived median line and show the interpretation of those tolerances.
▼ Explain and calculate virtual condition for a regular feature of size that has a form tolerance applied to it.
▼ Apply flatness to control a surface and show an interpretation of the flatness tolerance zone.
▼ Apply a flatness tolerance for a median plane and show an interpretation of the flatness tolerance zone.
▼ Apply circularity tolerances and show an interpretation of a circularity tolerance zone.
▼ Apply a cylindricity tolerance and show an interpretation of the cylindricity tolerance zone.

Review Exercises

Place your answers in the spaces provided. Accurately complete any required sketches. Show all calculations for problems that require mathematical solutions.

Multiple Choice

_____ 1. Allowable variations in the shape of an individual feature may be specified by size and _____ tolerances.
A. position
B. orientation
C. location
D. form

_____   2. Form variations on a feature of size may not exceed the _____ tolerance.
    A. position
    B. size
    C. orientation
    D. None of the above.

_____   3. A form tolerance controls _____ feature(s).
    A. One
    B. Two
    C. Three
    D. Any desired number of

_____   4. Generally, form tolerances applied on a surface _____ is permitted by the size tolerance.
    A. permits more variation of form than
    B. permits less variation of form than
    C. are equal to what
    D. Both A and C.

_____   5. Form tolerances are never _____.
    A. applied to an individual feature
    B. smaller than size tolerances
    C. used to establish location from datums
    D. specified with datum references

_____   6. _____ of ASME Y14.5 defines the assumption regarding material condition modifiers on form tolerances.
    A. Rule #1
    B. Rule #2
    C. Both A and B.
    D. Appendix A

_____   7. The least material condition for a hole is the _____.
    A. maximum allowable diameter
    B. minimum allowable diameter
    C. actual produced size
    D. None of the above.

_____   8. Two opposed sides of a rectangular part must be _____ when the part is at MMC.
    A. parallel
    B. flat
    C. coincident to the perfect form boundary
    D. All of the above.

_____   9. Two opposed sides of a rectangular part must be _____ when the part is at LMC.
    A. parallel
    B. flat
    C. straight
    D. None of the above.

Name _____

_____ 10. Parts subject to _____ are not controlled by Rule #1.
A. free state variation
B. damage
C. mass production
D. None of the above.

_____ 11. Perfect form at MMC is not a requirement when a straightness tolerance is applied to define allowable variation of _____.
A. a flat surface
B. the derived median line of a cylinder
C. surface elements on a cylinder
D. None of the above.

_____ 12. A specified derived median line straightness tolerance on a shaft _____.
A. also establishes a direct control of surface straightness
B. has no direct effect on surface straightness
C. must be specified in a special manner to also establish a tolerance for the surface straightness
D. None of the above.

_____ 13. If exception to Rule #1 is allowed on a feature, then a _____ must be applied on that feature.
A. small size tolerance
B. form tolerance
C. surface finish specification
D. None of the above.

_____ 14. A straightness tolerance used to specify allowable derived median line variation for a cylinder must include _____.
A. an MMC modifier
B. no modifier
C. a diameter symbol
D. None of the above.

_____ 15. Departure from MMC does not result in any change in the allowable form tolerance if the tolerance is specified to apply at _____.
A. MMC
B. RFS
C. LMC
D. All of the above.

_____ 16. The virtual condition of a hole is calculated by _____ the MMC size and straightness tolerance value.
A. finding the difference between
B. adding
C. multiplying
D. None of the above.

_____ 17. Functional gage sizes are based on the _____ of the features to be checked
if the feature has a size tolerance and a form tolerance applied at MMC.
A. LMC
B. MMC
C. virtual condition
D. nominal size

_____ 18. The flatness tolerance zone boundary may be at _____ orientation(s) to
the part.
A. only one defined
B. one of several defined
C. any
D. Either A or C.

_____ 19. A surface that has a flatness tolerance applied to it _____.
A. must also remain within the limits of size
B. may fall outside the limits of size by a value equal to the flatness
tolerance
C. must be oriented to the referenced datums
D. None of the above.

_____ 20. A circularity tolerance value is the _____ the boundary circles.
A. radial distance between
B. diameter difference between
C. center point offset for
D. None of the above.

_____ 21. Circularity _____ require surface location relative to the axis of the
controlled feature.
A. does
B. does not
C. may
D. combined with cylindricity may be used to

_____ 22. A cylindricity tolerance boundary is composed of two _____.
A. concentric circles
B. concentric cylinders
C. parallel planes
D. parallel lines

True/False

_____ 23. _True or False?_ Reducing size tolerance is one method of reducing allowable
form variations.

_____ 24. _True or False?_ It is preferable to reduce size tolerance to control form
rather than to apply a large size tolerance in combination with a small
form tolerance.

_____ 25. _True or False?_ Straightness tolerances applied on the surface line elements
of a shaft have the same effect as when the tolerance is applied to the
shaft diameter.

Name _____

_____ 26. *True or False?* A form tolerance specification that is applied to one flat surface will also be applicable to any surface that is parallel to the toleranced surface.

_____ 27. *True or False?* Stock materials, such as sheet and plate, must meet the requirements of Rule #1.

_____ 28. *True or False?* Straightness tolerances are never used to specify derived median line straightness for a shaft.

_____ 29. *True or False?* A straightness tolerance may be used to establish allowable variation for surface elements on a cone.

_____ 30. *True or False?* A derived median line straightness tolerance may be larger than the size tolerance.

_____ 31. *True or False?* Exception to the perfect form boundary requirements created by the size limits is never permitted regardless of the form tolerance values.

_____ 32. *True or False?* Functional gages may be used to inspect parts that have tolerances specified with the MMC modifier.

_____ 33. *True or False?* Unit length tolerances for derived median line straightness must be specified with a unit length of one inch.

_____ 34. *True or False?* Flatness tolerance feature control frames never include datum references to establish orientation requirements for the toleranced features.

_____ 35. *True or False?* A flatness tolerance that is attached to one surface establishes a requirement for that surface plus any other parallel surface.

_____ 36. *True or False?* Flatness of a derived median plane may only be specified by applying two flatness tolerances, one on each of the two surfaces that establish the derived median plane.

_____ 37. *True or False?* Circularity tolerances may be applied to any feature with a circular cross section.

_____ 38. *True or False?* Rule #1 in ASME Y14.5 defines what is often referred to as the *envelope principle*.

Fill in the Blank

_____ 39. The four form tolerances are straightness, flatness, circularity, and _____.

_____ 40. All form tolerances are specified in a _____ control frame.

_____ 41. Unless shown otherwise, the material condition modifier on a form tolerance is assumed to be _____.

_____ 42. A hole specification of .375″ ±.005″ diameter results in a perfect form boundary of _____ diameter.

_____ 43. A _____ tolerance specifies how close to perfectly straight a feature must be made.

_____ 44. A straightness tolerance applied to a feature of size is assumed to apply with the _____ modifier unless shown otherwise.

_____ 45. The virtual condition for a .375″ ±.003″ diameter shaft with a derived median line straightness tolerance of .007″ diameter is _____.

_____ 46. The MMC modifier indicates that the specified tolerance value may _____ as the toleranced feature departs from the MMC size.

_____ 47. Additional tolerance gained due to specification of the MMC modifier and departure of a feature from MMC is known as _____ tolerance.

_____ 48. Two parallel _____ bound the tolerance zone for a flatness tolerance.

_____ 49. Two _____ circles are the tolerance zone boundaries for a circularity tolerance.

_____ 50. _____ tolerances simultaneously establish requirements for circularity and straightness of cylindrical surfaces.

Short Answer

51. How are form variations on an individual feature specified to a value less than the size tolerance?

_____

_____

52. List the form tolerance categories.

_____

_____

53. A material condition modifier is applicable to the tolerance value when a form tolerance is applied to what type of feature?

_____

_____

54. Explain the difference between a surface and a feature of size.

_____

_____

Name _____

55. Define *maximum material condition*.

_____

_____

56. When all features on a part are at MMC, why is it possible for two adjacent features of size to be at an imperfect angle to one another?

_____

_____

57. Describe *free state variation*.

_____

_____

58. Describe how an exception to Rule #1 may be specified for a single feature.

_____

_____

59. Define *virtual condition*.

_____

_____

60. Explain the difference between a straightness tolerance specified on a flat surface and a flatness tolerance applied to the same surface.

_____

_____

Application Problems

All application problems are to be completed using correct dimensioning techniques. Show any required calculations.

61. If the bottom surface of a part produced to the given drawing is perfectly flat, what is the maximum possible flatness variation on the top surface?

.437±.010

_____

62. Apply a straightness tolerance of .007″ to control the straightness of surface elements on the given shaft.

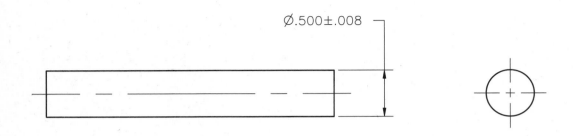

63. Apply a straightness tolerance of .007″ at MMC to specify a derived median line (axis) straightness on the given shaft or explain why it cannot be done.

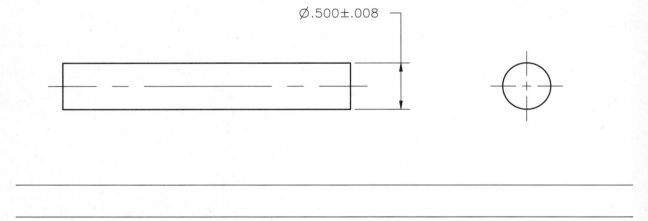

_____

_____

Name _____

64. Complete the interpretation drawing for the specified tolerances. Add any required tolerance zone boundaries, dimensions, or notes needed to complete the interpretation.

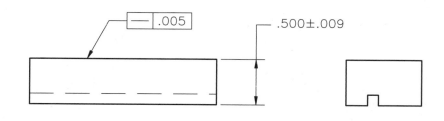

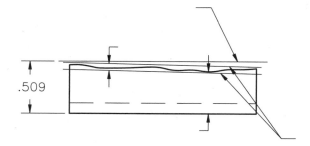
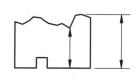

One cross section lengthwise
through the part

One cross section
across the part

65. Number the surfaces on the given part and enter the total number of surfaces on the blank provided.

66. Apply measurement values on the given part to illustrate the worst-case scenario that is allowed when both features are at MMC.

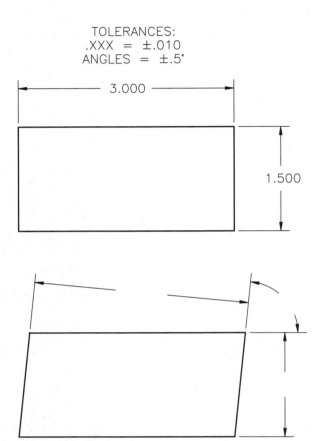

67. Show two methods of applying a straightness tolerance of .008″ on the bottom surface of the given view. Also show a thickness dimension of .750″ ±.015″.

68. Complete a straightness tolerance specification of .008″ diameter at MMC.

Name _____

69. Complete the interpretation drawing for the specified tolerances. Add any required tolerance zone boundaries, dimensions, or notes needed to complete the interpretation.

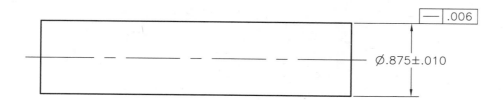

Ø.875±.010

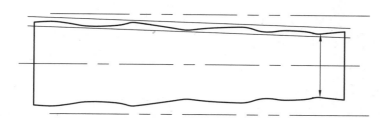

70. Complete the interpretation drawing for the specified tolerances. Add any required tolerance zone boundaries, dimensions, or notes needed to complete the interpretation.

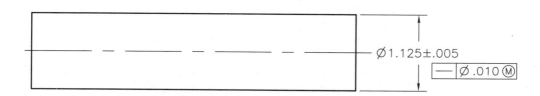

Ø1.125±.005

71.   What is the virtual condition for the hole? _____

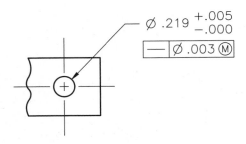

72.   Apply a straightness tolerance specification that results in a virtual condition of .216″ diameter.

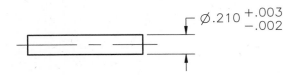

73.   Apply a straightness tolerance specification to achieve overall length derived median line straightness of .015″ diameter at MMC and unit length derived median line straightness of .005″ diameter at MMC per 1.00″ of length.

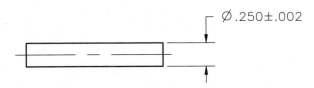

74.   Sketch a gage to check the unit length specification in the given figure. Apply dimensions to show the theoretical dimensions for a perfect gage. Do not apply gage tolerances.

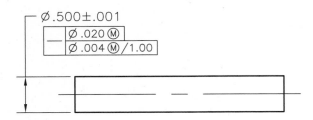

Name _____

75. Show two methods of applying a flatness tolerance of .010″ on one of the large surfaces on the part in the following illustration.

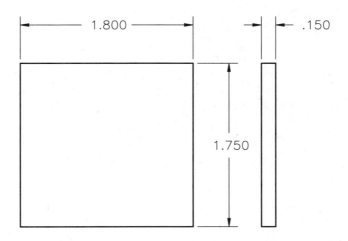

76. Complete the interpretation drawing for the specified tolerances. Add any required tolerance zone boundaries, dimensions, or notes needed to complete the interpretation.

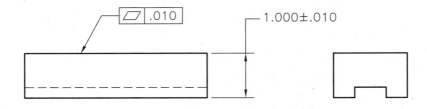

77. Draw a feature control frame that establishes an overall flatness tolerance of .020″ and a unit area flatness of .009″ per square inch.

78. Apply a circularity tolerance that permits .010″ surface variation when measured radially from a perfect circumscribing circle.

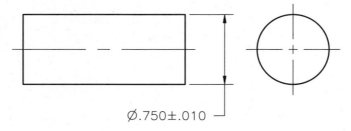

Ø.750±.010

79. Complete the interpretation drawing for the specified tolerances. Add any required tolerance zone boundaries, dimensions, or notes needed to complete the interpretation.

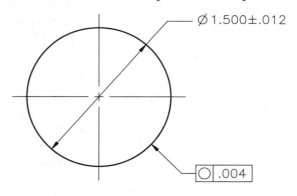

Ø 1.500±.012

○ | .004

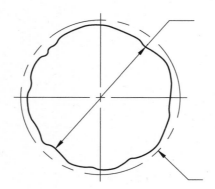

Name _____

80. Apply a tolerance specification that requires surface conditions to fall within two concentric cylinders separated by .005″.

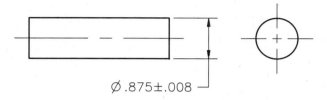

Ø .875±.008

81. A shaft is produced at a diameter of .559″. The specified size is .562″ ±.004″ and a derived median line straightness tolerance of .003″ diameter at MMC is specified. What is the allowable straightness variation on the produced part?

_____

NOTES

Chapter 6

Datums and Datum Feature References

Name _____   Date_____   Class _____

Reading

Read Chapter 6 of the *GD&T: Application and Interpretation* textbook prior to completing the review exercises.

Objectives

A combination of activities is required to achieve the following objectives. Completing the reading assignment and the following review exercises is an important part of achieving the objectives. Familiarization with the objectives prior to completion of the reading assignment and review exercises will make mastery of the objectives easier. After completing the reading assignment and completing the review exercises, you will be able to:

▼ Define the difference between a theoretically perfect datum and a datum feature.
▼ Explain how to create a datum reference frame through references made in a feature control frame.
▼ Utilize all methods for identifying datum features, including the use of target points, lines, and areas.
▼ Make datum feature references in a feature control frame using the correct order of precedence.
▼ Explain how a datum reference frame may be simulated when three mutually perpendicular surfaces are referenced as datum features.
▼ Use material boundary modifiers on datum feature references and explain the significance of the modifiers.
▼ Identify the degrees of freedom constrained by each referenced datum feature in a datum reference frame.

Review Exercises

Place your answers in the spaces provided. Accurately complete any required sketches. Show all calculations for problems that require mathematical solutions.

Multiple Choice

_____   1. Datum feature references may be contained in a _____.
    A. datum reference frame
    B. feature control frame
    C. datum system
    D. machine part

_____   2. A tolerance specification shown in a feature control frame may include _____ datum reference(s).
　　　　　　　　　　　　　　　A. one
　　　　　　　　　　　　　　　B. two
　　　　　　　　　　　　　　　C. three
　　　　　　　　　　　　　　　D. Any of the above.

_____   3. The first datum feature reference in a tolerance specification identifies the _____ datum reference.
　　　　　　　　　　　　　　　A. primary
　　　　　　　　　　　　　　　B. secondary
　　　　　　　　　　　　　　　C. tertiary
　　　　　　　　　　　　　　　D. None of the above.

_____   4. Planes in a datum reference frame are always _____.
　　　　　　　　　　　　　　　A. perfect
　　　　　　　　　　　　　　　B. mutually perpendicular
　　　　　　　　　　　　　　　C. Both A and B.
　　　　　　　　　　　　　　　D. Neither A nor B.

_____   5. The factor that is irrelevant when defining datum feature references for a tolerance specification is _____.
　　　　　　　　　　　　　　　A. functional requirements
　　　　　　　　　　　　　　　B. fabrication methods
　　　　　　　　　　　　　　　C. inspection methods
　　　　　　　　　　　　　　　D. alphabetical order of datum letters

_____   6. The datum target symbol is used to identify datum _____.
　　　　　　　　　　　　　　　A. targets
　　　　　　　　　　　　　　　B. features
　　　　　　　　　　　　　　　C. planes
　　　　　　　　　　　　　　　D. axes

_____   7. A datum target symbol is a circle with a _____ line across it.
　　　　　　　　　　　　　　　A. vertical
　　　　　　　　　　　　　　　B. horizontal
　　　　　　　　　　　　　　　C. diagonal
　　　　　　　　　　　　　　　D. Both A and B.

_____   8. A surface plate or other tooling device used to contact a datum feature acts as a datum _____.
　　　　　　　　　　　　　　　A. plane
　　　　　　　　　　　　　　　B. simulator
　　　　　　　　　　　　　　　C. axis
　　　　　　　　　　　　　　　D. reference frame

_____   9. A primary reference to a cylindrical datum feature establishes a _____.
　　　　　　　　　　　　　　　A. datum axis
　　　　　　　　　　　　　　　B. datum plane
　　　　　　　　　　　　　　　C. coordinate system
　　　　　　　　　　　　　　　D. centerline

Name _____

_____ 10. A _____ leader extending from a datum target symbol to a datum target
indicates the target is on the far side of the object.
   A. solid
   B. dashed
   C. phantom
   D. None of the above.

_____ 11. Single point contact at a target point can be achieved with a _____.
   A. side of a round dowel
   B. chuck or collet
   C. spherical-ended tool post
   D. All of the above.

_____ 12. An end view of a _____ is shown with the same symbol as a target point.
   A. target line
   B. target area
   C. datum surface
   D. None of the above.

_____ 13. Target areas have a _____ shape.
   A. round
   B. square
   C. rectangular
   D. Any of the above.

_____ 14. _____ datum reference frame(s) is/are created if one feature control
frame references datum A primary, B secondary, and C tertiary; and
another feature control frame references datum B primary, C secondary,
and A tertiary.
   A. One
   B. Two
   C. Three
   D. All of the above.

_____ 15. A flat surface on a part will stabilize on _____ point(s) or more when set
on a surface plate.
   A. one
   B. two
   C. three
   D. None of the above.

_____ 16. Datum _____ is a means of approximating the theoretical location of
the datums.
   A. referencing
   B. identification
   C. targeting
   D. simulation

17. Identifying a hole as a datum feature is a means of establishing a _____.
    A. datum axis
    B. datum plane
    C. datum target
    D. virtual hole

18. A datum feature surface may be identified by a datum feature symbol placed _____.
    A. on an extension line from the surface
    B. on the feature
    C. on a feature control frame attached to the feature
    D. All of the above.

19. A reference to datum A primary, B secondary, and C tertiary creates _____ a reference to datum A primary, C secondary, and B tertiary.
    A. the same datum reference frame as
    B. the same coordinate system as
    C. a different datum reference frame than
    D. None of the above.

20. Multiple groups of features are assumed to _____ if the tolerance specifications on the groups reference the same datums in the same order of precedence and include the same material condition modifiers.
    A. create one pattern (simultaneous requirements)
    B. create multiple patterns (separate requirements)
    C. create confusion
    D. Both B and C.

21. There must be at least _____ target point(s) identified for a flat surface that is referenced as a primary datum.
    A. one
    B. two
    C. three
    D. four

22. The distance between stepped datum targets is defined with _____.
    A. basic dimensions
    B. limit dimensions
    C. plus or minus tolerances
    D. None of the above.

23. _____ targets are used to establish a datum plane by contacting features in a manner that causes the feature to center.
    A. Equalizing
    B. Small
    C. Large
    D. Stepped

True/False

24. *True or False?* Datum features are typically identified by attaching symbols to centerlines and other theoretical entities.

25. *True or False?* Tolerance specifications that reference datum features require that measurements be verified relative to the datums rather than to the imperfect part surfaces.

Name _____

_____ 26. *True or False?* The letter used for a primary datum feature reference must precede the letter in the alphabet used for a secondary datum reference.

_____ 27. *True or False?* Using implied datums is permitted since this practice saves time.

_____ 28. *True or False?* A datum target point shown on a drawing indicates that the target location is intended to make point contact with the tooling.

_____ 29. *True or False?* Contact with a datum target line on a flat surface may be achieved by contacting the side of a dowel pin.

_____ 30. *True or False?* The perimeter of a target area must always be shown with a phantom line.

_____ 31. *True or False?* Datum precedence shown in a feature control frame affects how the datum features are used to establish a datum reference frame.

_____ 32. *True or False?* A secondary datum feature that is produced with an angular variation relative to the primary datum feature causes the datum reference frame to be distorted.

_____ 33. *True or False?* The minimum number of points on a flat surface that must make contact to establish a secondary datum plane is two.

_____ 34. *True or False?* A datum feature triangle should not be attached to a dimension line.

_____ 35. *True or False?* Before a means of datum simulation can be determined, it is necessary to know the order of precedence of all datums and the material boundary modifier applicable to each reference.

_____ 36. *True or False?* A datum feature cannot be referenced as a primary datum in one specification and as a secondary datum in another specification.

_____ 37. *True or False?* Multiple (compound) datum feature references separated by a dash create a requirement to use the identified features to establish one datum.

_____ 38. *True or False?* ASME Y14.5 specifies that datum feature symbols should not be shown on centerlines.

_____ 39. *True or False?* Datum targets are permitted on cylindrical features such as holes and shafts.

_____ 40. *True or False?* More than three datum targets may be placed on a single datum feature.

_____ 41. *True or False?* It is a poor practice to combine datum target areas and datum target points on the same datum feature.

Fill in the Blank

_____ 42. A datum reference frame made up of three mutually perpendicular planes may be established by referencing _____ datum feature(s) that are flat surfaces.

_____ 43. A _____ feature symbol is used to identify a surface or feature of size as a datum feature.

_____ 44. A datum _____ is established by a flat surface that is identified as a datum feature.

_____ 45. A(n) _____ line (_type of line_) is normally used to show the perimeter of a datum target area.

_____ 46. A primary datum feature establishes location of the first plane in the datum _____ frame.

_____ 47. _____ points are required to define a plane.

_____ 48. The secondary datum plane in a datum reference frame must be oriented _____ to the primary plane.

_____ 49. _____ flat surfaces must be referenced to establish three planes in a datum reference frame.

_____ 50. The diameter of a round target area may be shown in the _____ half of the datum target symbol.

_____ 51. The order of datum _____ shown in a feature control frame must be considered when defining datum targets on a drawing.

_____ 52. If a primary datum plane is established by a flat surface, _____ holes must be referenced as datum features to completely establish and clock the datum reference frame.

_____ 53. Surfaces that lie in more than one plane are called _____ datum surfaces when they are used to establish one datum plane.

_____ 54. If a primary datum feature reference is to a datum feature of size and the reference includes the MMB modifier, then the datum simulator size is equal to the _____ of the datum feature.

_____ 55. If a secondary datum feature reference is to a datum feature of size and the reference includes the MMB modifier, then the datum simulator size is equal to the _____ of the datum feature.

Name _____

Short Answer

56. What is the difference between a datum feature and a datum?

_____

_____

57. List two types of tolerance specifications that typically include datum feature references.

_____

_____

58. State one reason why it is preferable to measure from a datum reference frame rather than from datum features.

_____

_____

59. Explain why it is ambiguous to place a datum feature symbol on the centerline of a counterbored hole.

_____

_____

60. Describe two of the methods for applying a datum feature symbol to indicate that a flat surface is a datum feature.

_____

_____

61. List the three types of datum targets.

_____

_____

62. Explain why at least three target points are needed on a surface that is referenced as a primary datum feature.

_____

_____

63. List one factor that should be considered when determining the size of a datum target area and explain why the factor should be considered.

_____

_____

64. If a workpiece is considered unstable on the primary datum simulator, what may be done?

_____

_____

65. What is the result of applying a datum feature symbol to the width dimension on a slot?

_____

_____

66. When are material boundary modifiers applicable on datum feature references?

_____

_____

67. What is the difference between datum reference A-B and AB?

_____

_____

Name _____

Application Problems

All application problems are to be completed using correct dimensioning techniques. Show any required calculations.

68. Identify the order of precedence for each of the datum feature references.

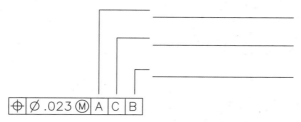

69. In the drawing, identify a datum feature reference and a datum feature. In the interpretation view, identify a datum feature and a datum plane.

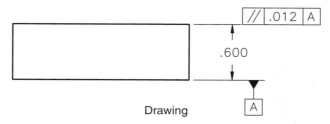

Drawing

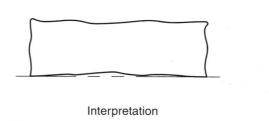

Interpretation

70. A sketch of a manufactured part is given below the drawing. Sketch the planes of a datum reference frame in all the views of the manufactured part. Label each of the datum planes that make up the datum reference frame and note the number of points of contact required with the feature surface. Complete the sketch assuming that a feature control frame references datum feature A primary, B secondary, and C tertiary.

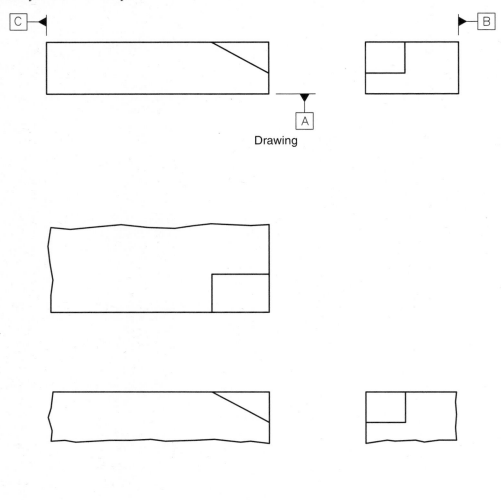

71. Complete the datum target symbol for target location A3. It is a target area with a .50" diameter.

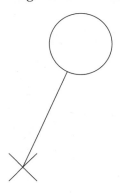

Name _____

72. Identify the diameter of the given cylinder as datum feature A so that a datum axis is established.

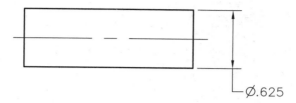

∅.625

73. Show datum target points on the given part. Use a number of targets that permit datum references in the order of precedence shown in the given drawing. Label all targets. Use basic dimensions to define target locations.

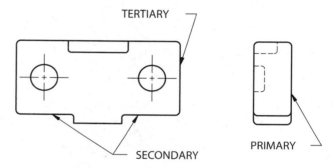

TERTIARY

SECONDARY

PRIMARY

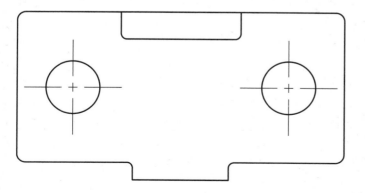

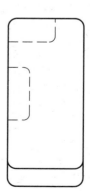

74. Identify each of the given target symbols.

✕   A. _____

    C. _____

— — - - —   B. _____

75. Complete the given feature control frame. Reference datum feature D primary, B secondary, and C tertiary.

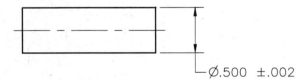

76. Identify the shaft diameter as datum feature A and the right end as datum feature B.

Ø.500 ±.002

Name _____

77. A drawing and manufactured part are given. Sketch a tool that properly locates the datum reference frame for the manufactured part. Show possible points of contact that would stabilize the part in the tool and meet the order of precedence for the referenced datums (Hint: Show at least three points on the primary datum feature).

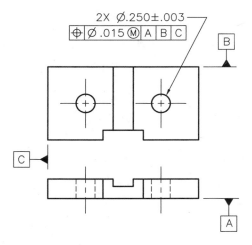

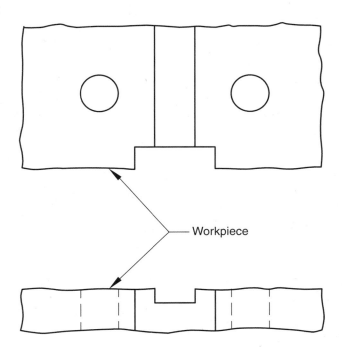

Workpiece

78. Sketch and dimension the gage features required to establish the datum reference frame for the shown part. Superimpose the gage on the given views.

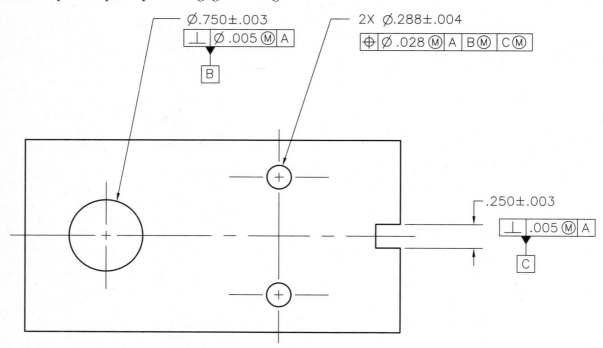

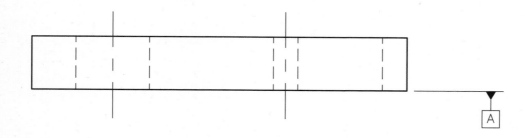

Name _____

79. Assume the bottom surface of the shown part is referenced in two feature control frames. It is referenced as primary datum A in one specification. It is referenced as secondary datum E in another specification. Show targets that permit the two datum references.

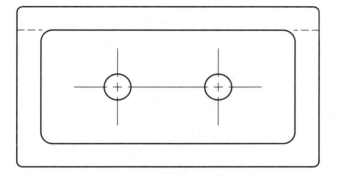 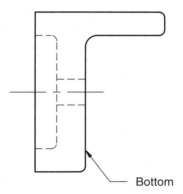

Bottom

80. Identify the centerdrill countersinks as datum features A and B. Complete the total runout specification by showing a primary datum reference to compound datum features A and B.

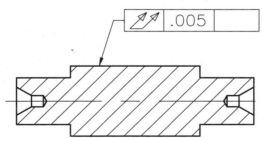

81. Explain why the shown drawing is wrong and correct the drawing.

_____

_____

_____

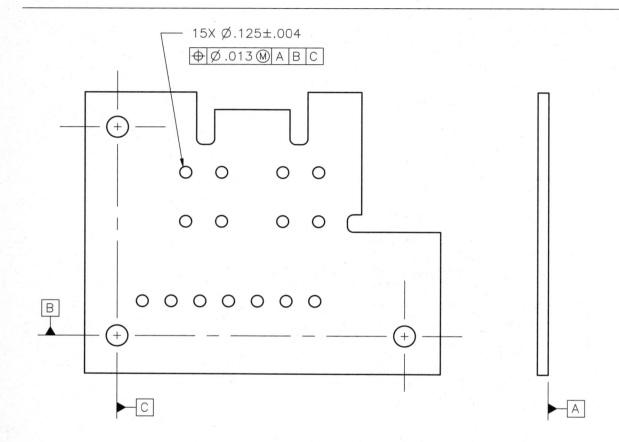

Name _____

82. Sketch the datum simulators required for the given part. Apply nominal size and location dimensions for the simulators. Do not superimpose the sketch on the given part.

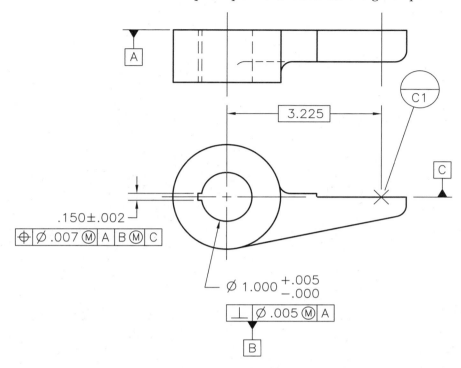

83. The given part drawing shows a front and bottom view with dimensions. A front and top view are shown in phantom lines. Sketch the datum simulators required for the given part in the views where the part is shown in phantom. Apply nominal location dimensions for the target point locators.

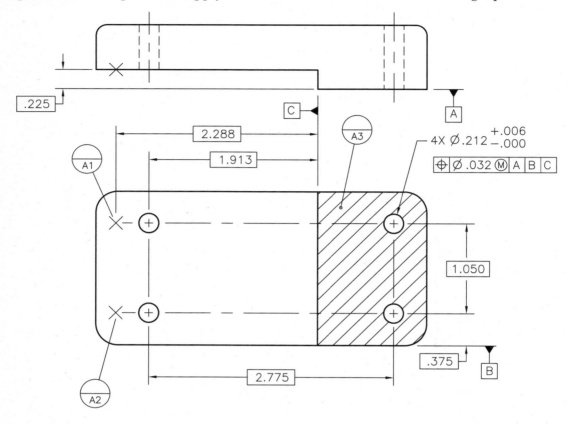

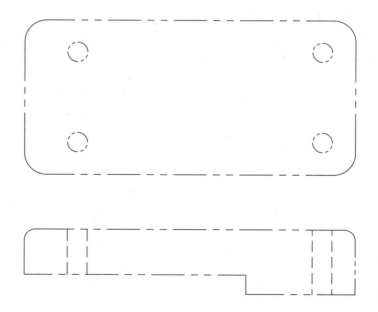

Name _____

84. Complete the interpretation drawing for the one dimension that has note #1 applied to it. Show the datums and the dimensions to the tolerance zone for the one dimensioned feature that is affected.

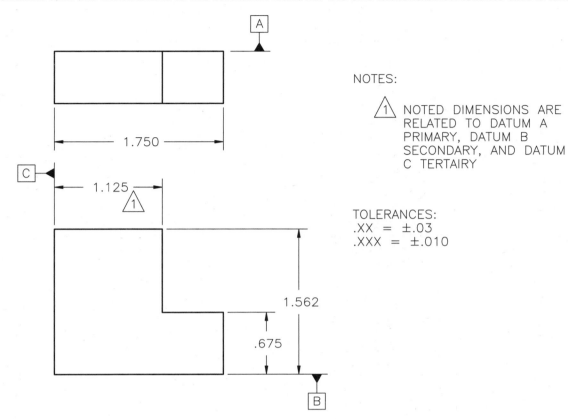

NOTES:

△1 NOTED DIMENSIONS ARE RELATED TO DATUM A PRIMARY, DATUM B SECONDARY, AND DATUM C TERTAIRY

TOLERANCES:
.XX = ±.03
.XXX = ±.010

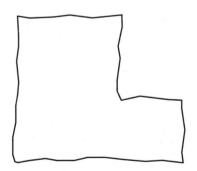

NOTES

Chapter 7
Orientation Tolerances

Name _____ Date _____ Class _____

Reading

Read Chapter 7 of the *GD&T: Application and Interpretation* textbook prior to completing the review exercises.

Objectives

A combination of activities is required to achieve the following objectives. Completing the reading assignment and the following review exercises is an important part of achieving the objectives. Familiarization with the objectives prior to completion of the reading assignment and review exercises will make mastery of the objectives easier. After completing the reading assignment and completing the review exercises, you will be able to:

▼ Identify, apply, and interpret orientation tolerances.
▼ Complete orientation tolerance specifications including one or two datum feature references.
▼ Explain the effects of material condition modifiers when orientation tolerances are applied to features of size.
▼ Calculate the virtual condition for internal and external features of size to which an orientation tolerance is applied.
▼ Complete tolerance specifications that include orientation and form requirements on a single feature.

Review Exercises

Place your answers in the spaces provided. Show all calculations for problems that require mathematical solutions.

Multiple Choice

_____ 1. There must be _____ datum feature reference(s) in a perpendicularity tolerance specification.
A. no
B. only one
C. one or more
D. two or more

_____   2. Two _____ form the tolerance zone boundary when an orientation
tolerance is applied to a flat surface.
A. intersecting lines
B. intersecting surfaces
C. parallel lines
D. parallel planes

_____   3. Which of the following is not an orientation tolerance?
A. Angularity
B. Concentricity
C. Parallelism
D. Perpendicularity

_____   4. The _____ condition caused by an orientation tolerance applied at MMC
on a hole is determined by subtracting the orientation tolerance from the
minimum size limit of the hole.
A. virtual
B. resultant
C. MMC
D. LMC

_____   5. A parallelism tolerance applied to a flat surface results in a tolerance zone
that is bounded by _____ that are parallel to a referenced datum plane.
A. lines
B. planes
C. cylinders
D. None of the above.

_____   6. Application of a parallelism tolerance on a hole requires that a _____ be
assumed or applied on the tolerance value.
A. minimum value
B. maximum value
C. metric value
D. material condition modifier

_____   7. A perpendicularity tolerance applied to a flat surface on the end of a
rectangular part defines an orientation requirement for _____.
A. only the surface to which it is applied
B. both the surface to which it is applied and the opposite end of
the part
C. the center plane of the toleranced feature of size
D. None of the above.

_____   8. A perpendicularity tolerance applied to the width dimension on a slot
establishes a requirement on _____ to a value equal to the tolerance value.
A. both sides of the slot
B. the side of the slot closest to the tolerance specification
C. the center plane of the slot
D. All of the above.

Name _____

_____ 9. An orientation tolerance applied to _____ may result in surface variation that lies outside the tolerance zone, but a plane tangent to the surface must be within the tolerance zone.
    A. an individual feature
    B. multiple features
    C. a unit area
    D. a tangent plane

True/False

_____ 10. *True or False?* Parallelism tolerances may only be applied to flat surfaces.

_____ 11. *True or False?* An orientation tolerance may be used to establish a location requirement.

_____ 12. *True or False?* An orientation tolerance should not be applied to a feature that already has another tolerance type such as a position tolerance.

_____ 13. *True or False?* An orientation tolerance including the MMC modifier and applied to an internal feature of size, such as a hole, creates a virtual condition that is smaller than the MMC size of the toleranced feature.

_____ 14. *True or False?* A parallelism tolerance establishes an orientation requirement, and does not establish the maximum and minimum limits of size for a feature.

_____ 15. *True or False?* A parallelism tolerance of .008″ can define the allowable variation on the distance (location or size) between two flat surfaces.

_____ 16. *True or False?* A diameter symbol is needed when a parallelism tolerance is applied to control the parallelism of the axis for one hole to the axis of another hole.

_____ 17. *True or False?* Ninety degree angles in an orthographic view do not require dimensions to show the angle.

_____ 18. *True or False?* A perpendicularity tolerance must never reference two datum features.

_____ 19. *True or False?* A secondary datum feature reference in a perpendicularity tolerance specification stops rotation of the part on the primary datum.

Fill in the Blank

_____ 20. _____ tolerances are used to specify angularity, parallelism, and perpendicularity requirements relative to one or more datums.

_____ 21. _____ tolerance provides control of a flat surface at any angle.

_____ 22. When no material condition modifier is shown on an orientation tolerance, the _____ material condition modifier is assumed to apply.

_____ 23. A parallelism tolerance value applied to a flat surface must not be _____ than the tolerance value that locates the surface.

_____ 24. The primary datum feature referenced in a perpendicularity tolerance specification must be at a(n) _____ angle to the toleranced feature.

_____ 25. A produced surface with an orientation tolerance applied to it may have form variation that is equal to or _____ than the orientation tolerance.

_____ 26. An angularity tolerance specification applied to a flat surface results in a tolerance zone bounded by two _____.

_____ 27. The angle dimension value must be _____ when an angularity tolerance is applied.

Short Answer

28. List the three orientation tolerances.

_____

_____

29. When is a material condition modifier applicable to an orientation tolerance?

_____

_____

30. Describe what is meant by the term *virtual condition* when the term is associated with a shaft.

_____

_____

31. How much parallelism variation may exist when the size dimension between two surfaces is ±.015″?

_____

32. Explain why it is possible to have a location tolerance of .050″ between two holes and a parallelism tolerance of .010″ between the same two holes.

_____

_____

_____

Name _____

33. When is a 90° angle understood to be basic?

_____

_____

34. Determine the virtual condition for a .563", plus .005", minus .000" diameter pin that has a .012" diameter perpendicularity tolerance.

_____

35. Determine the virtual condition for a .750", plus .006", minus .002" diameter hole that has a .010" diameter perpendicularity tolerance.

_____

Application Problems

All application problems are to be completed using correct dimensioning techniques. Show any required calculations.

36. Show the tolerance zone for each of the inclined surfaces.

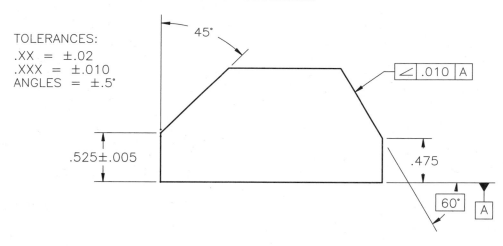

TOLERANCES:
.XX = ±.02
.XXX = ±.010
ANGLES = ±.5°

37. Identify each of the shown symbols.

∠ _____

⊥ _____

// _____

38. Complete a feature control frame that controls a flat surface to be perpendicular to datum surface A within a zone that is .006" wide.

39. Calculate the virtual condition for the shown hole.

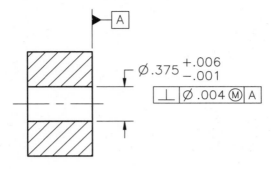

40. Calculate the virtual condition for the shown pin.

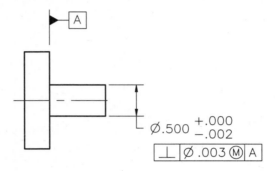

41. Surface B must be parallel within .005″ to a datum established by surface A. Surface C must be parallel within .010″ to the same datum. Show all required tolerance specifications.

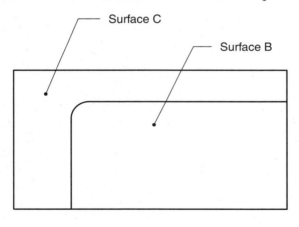

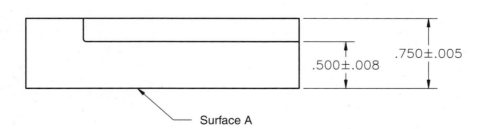

Name _____

42. Complete the interpretation drawing and show the allowable tolerance zones for all specified tolerances.

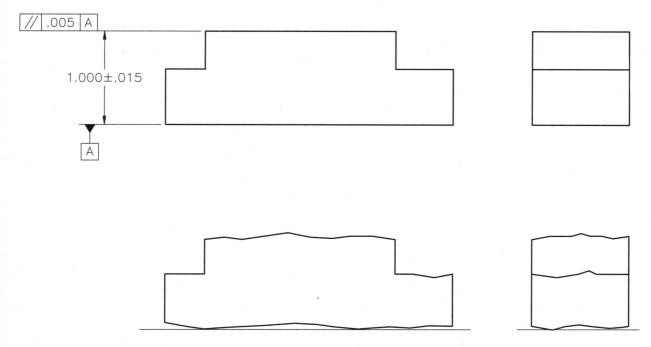

43. Apply a size dimension and tolerance to permit the slot width to vary by .020″ total and apply geometric tolerance(s) to require the sides of the slot to be parallel to one another within .008″.

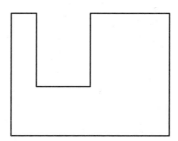

44. Apply a location tolerance of ±.025″ between the shown holes. Establish one hole as a datum feature. Apply a tolerance that defines a parallelism requirement between the holes to .010″ when both holes are at MMC. There is more than one acceptable solution.

Hole #1 = Datum hole
Hole #2 = Controlled hole
Hole diameter = .250±.003

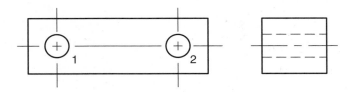

Name _____

45. Complete an interpretation drawing that shows the permitted perpendicularity tolerance zone.

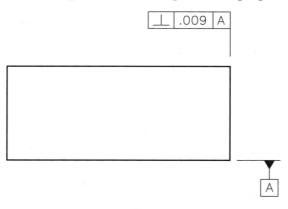

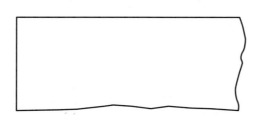

46. Complete an interpretation drawing that shows the permitted perpendicularity tolerance zone.

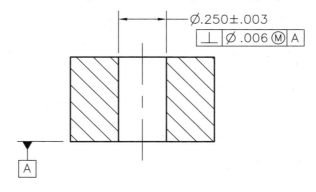

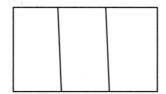

47. A hole size specification and perpendicularity tolerance is shown. Complete the given table to show each permitted hole size and show the corresponding allowable perpendicularity tolerances.

Given hole specification

$\varnothing.375 \begin{smallmatrix} +.004 \\ -.001 \end{smallmatrix}$

| ⊥ | $\varnothing$ .007 Ⓜ | A |

Produced hole diameter	Allowable perpendicular tolerance
.374	_____
.375	_____
.376	_____
.377	_____
.378	_____
.379	_____

48. Apply a perpendicularity tolerance that results in a virtual condition of .379″ diameter for the pin.

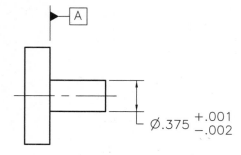

49. Apply a perpendicularity tolerance that results in a virtual condition of .379″ diameter for the hole.

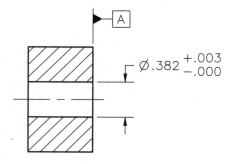

Name _____

50. Complete an interpretation drawing that shows the permitted angularity tolerance zone.

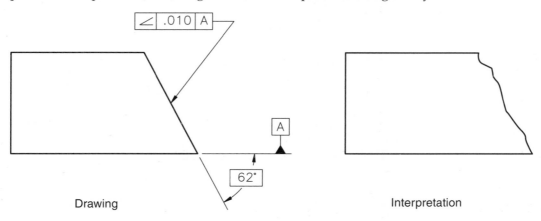

Drawing   Interpretation

51. Complete an interpretation drawing that shows the permitted angularity tolerance zone. Show a permissible surface condition that lies partially outside the specified tolerance zone.

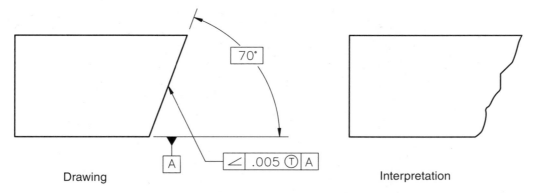

Drawing   Interpretation

52. Complete a feature control frame that controls parallelism of the top surface to .015″ relative to datum feature A and flatness to .005″.

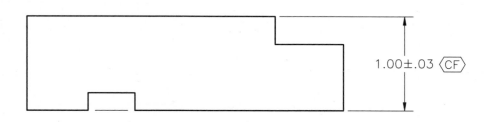

53. Complete a feature control frame that controls perpendicularity of the hole to .012″ at MMC relative to datum feature A and axis straightness to .004″ at MMC.

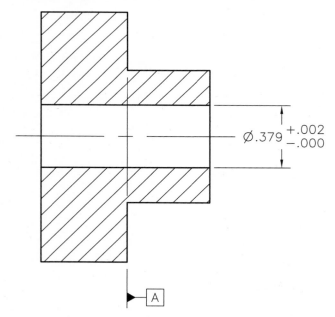

Chapter 8

Position Tolerancing Fundamentals

Name _____   Date _____   Class _____

Reading

Read Chapter 8 of the *GD&T: Application and Interpretation* textbook prior to completing the review exercises.

Objectives

A combination of activities is required to achieve the following objectives. Completing the reading assignment and the following review exercises is an important part of achieving the objectives. Familiarization with the objectives prior to completion of the reading assignment and review exercises will make mastery of the objectives easier. After completing the reading assignment and completing the review exercises, you will be able to:

▼ Complete feature control frames for position tolerances, properly using the diameter symbol, material condition modifiers, and datum references.
▼ Sketch the proper location and shape for position tolerance zones.
▼ Describe the effect of an MMC, LMC, or RFS modifier on a position tolerance.
▼ Provide examples that prove the validity of the MMC concept as it applies to position tolerances.
▼ Calculate position tolerances for simple fixed and floating fastener conditions.
▼ Calculate the allowable bonus tolerance for a produced part on which a position tolerance is specified at MMC.
▼ Use calculation techniques to verify whether produced hole locations meet specified drawing tolerances.
▼ Cite advantages of position tolerances when compared to coordinate hole location tolerances.

Review Exercises

Place your answers in the spaces provided. Accurately complete any required sketches. Show all calculations for problems that require mathematical solutions.

Multiple Choice

_____   1. Location dimensions must be _____ if a position tolerance is applied to the located feature.
   A. nominal values
   B. limit values
   C. toleranced
   D. basic

2. Application of position tolerances for hole locations requires that datum _____ be identified on the part.
    A. features
    B. planes
    C. axes
    D. None of the above.

3. The 1982 and later issues of the dimensioning and tolerancing standard prohibit _____ on position tolerances.
    A. implied datums
    B. the use of MMC
    C. Both A and B.
    D. Neither A nor B.

4. Rule #2 requires that material condition modifiers be shown on position tolerances when _____ applies.
    A. MMC
    B. LMC
    C. RFS
    D. Either A or B.

5. A position tolerance zone for a round hole is normally _____ in shape.
    A. conical
    B. cylindrical
    C. elliptical
    D. square

6. The _____ modifier indicates that a position tolerance may increase as a hole size departs from the minimum permitted diameter.
    A. MMC
    B. LMC
    C. RFS
    D. None of the above.

7. If two mating parts each have clearance holes through which a bolt is inserted, a _____ condition exists.
    A. least material
    B. floating fastener
    C. maximum material
    D. fixed fastener

8. Fastener and clearance hole _____ are used to calculate position tolerances that will achieve fastener installation.
    A. nominal sizes
    B. maximum size limits
    C. least material conditions
    D. maximum material conditions

9. Specification of a position tolerance with an MMC modifier results in a(n) _____ tolerance when the feature is produced at any allowable size other than MMC.
    A. undefined
    B. bonus
    C. reduced
    D. None of the above.

Name _____

_____ 10. The allowable position tolerance is equal to the sum of the _____ and the bonus tolerance.
   A. specified tolerance
   B. feature size tolerance
   C. specified feature size
   D. actual produced diameter

_____ 11. Specified hole limits of .384″ MIN and .394″ MAX are given. A position tolerance of .009″ diameter at MMC is specified for the hole. What is the allowable position tolerance for a hole produced at .386″ diameter?
   A. .007″
   B. .009″
   C. .011″
   D. .015″

_____ 12. A position tolerance referenced to three datum planes requires that all hole locations be measured from _____.
   A. the datum planes
   B. the datum features
   C. one another
   D. with a coordinate measurement machine

_____ 13. _____-shaped position tolerance zones permit the same amount of hole location variation in all directions.
   A. Round
   B. Square
   C. Rectangular
   D. None of the above.

_____ 14. A position tolerance applied to a thread defines the location requirement for the _____ diameter when no additional notation is provided.
   A. major
   B. pitch
   C. minor
   D. root

_____ 15. A _____ tolerance zone lies outside the toleranced feature.
   A. projected
   B. position
   C. runout
   D. bonus

_____ 16. _____ feature control frames may be applied to create a bidirectional tolerance on a slotted hole.
   A. Two
   B. Composite
   C. Combined
   D. None of the above.

_____ 17. A position tolerance applied to establish the location tolerance for a slot, such as a keyseat, requires that _____ of the slot be located within the allowable tolerance.
A. one side
B. one end
C. the center plane
D. All of the above.

True/False

_____ 18. *True or False?* Position tolerances are applied to features of size and bounded features.

_____ 19. *True or False?* Every position tolerance specification must include a material condition modifier symbol on the tolerance value.

_____ 20. *True or False?* Beginning with ANSI Y14.5M-1982, implied datums are not permitted on position tolerance specifications.

_____ 21. *True or False?* It is acceptable to show a material boundary modifier on a datum reference in a position tolerance specification if the datum feature is a feature of size.

_____ 22. *True or False?* The theoretical true position for a hole defined by a basic dimension means there is no position tolerance allowed and the hole must be perfectly located.

_____ 23. *True or False?* The allowable size of the position tolerance zone is dependent on the amount of hole size departure from MMC if no material condition modifier is shown in the position tolerance specification.

_____ 24. *True or False?* An MMC modifier on a position tolerance can permit greater freedom in how a part is produced.

_____ 25. *True or False?* T = H − F is a formula that may be used for a floating fastener condition in which both holes are the same size and the position tolerance applied to each hole is the same value.

_____ 26. *True or False?* If an MMC modifier is applied to a position tolerance on a hole, the allowable position tolerance increases as the hole size is increased.

_____ 27. *True or False?* Functional gages must be used to verify hole positions when position tolerances are specified.

_____ 28. *True or False?* Position tolerances permit utilization of the full amount of tolerance that is functionally possible for a hole, but coordinate tolerances do not.

_____ 29. *True or False?* Position tolerances are not appropriate or needed when the allowable variation is relatively large.

Name _____

_____ 30. *True or False?* Square tolerance zones do not permit the same amount of permissible hole location variation in all directions relative to the nominal position.

_____ 31. *True or False?* Bonus tolerances may be utilized when coordinate tolerances are applied to hole locations.

Fill in the Blank

_____diameter_____ 32. A(n) _____ symbol placed in front of the position tolerance value indicates the tolerance zone is round.

_____ 33. Position tolerance zones are centered on the _____ position defined by basic dimensions.

_____ 34. A hole for a press fit pin would typically have a position tolerance that applies at the _____ material condition.

_____ 35. A large amount of clearance between a hole and fastener permits _____ position tolerance than would be possible for a small amount of clearance.

_____ 36. The use of the _____ material condition results in no allowable change in the specified tolerance regardless of the produced feature size.

_____ 37. Concentric circles superimposed on a grid may be used to represent tolerance zone _____ when paper gaging.

_____ 38. A round tolerance zone has _____ percent more area than a square tolerance zone if the effect of bonus tolerances is ignored.

_____ 39. The letter *P* inside a circle indicates a requirement for a(n) _____ tolerance zone.

Short Answer

40. Why is it necessary to permit tolerances on the location of features?

_____

_____

41. Describe one method that may be used to show the number of holes to which a position tolerance applies.

_____

_____

_____

42. Describe one reason why implied datums should not be used, even when working to an old issue of the standard.

_____

_____

_____

43. List the two general fastener conditions for which position tolerances may be calculated.

_____

_____

44. Describe a fixed fastener condition.

_____

_____

45. What is the formula used to calculate the position tolerance for a fixed fastener condition? Assume even distribution of the allowable tolerance for the two parts.

_____

46. Coordinates specified for a hole are: X = 1.375″ and Y = 3.250″. A hole is produced at X = 1.381″ and Y = 3.248″. What is the diameter of position variation from true position? Show your calculations.

_____

_____

_____

_____

47. Explain why a functionally correct round tolerance zone has a diameter that circumscribes a calculated square tolerance zone.

_____

_____

_____

_____

Name _____

48. What is the effect on the hole and counterbore when a single position tolerance specification is applied to the hole and counterbore callout?

_____

_____

_____

49. Explain an advantage of bidirectional position tolerances applied at MMC as compared to plus or minus location tolerances on a hole.

_____

_____

_____

Application Problems

All application problems are to be completed using correct dimensioning techniques. Show any required calculations.

50. Identify a basic dimension, a datum feature symbol, and a position tolerance specification.

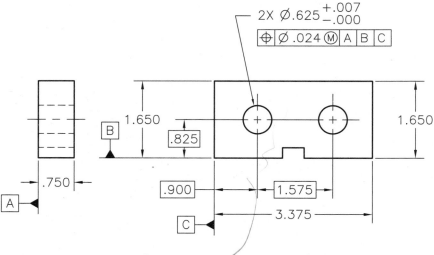

51. Complete a feature control frame for a position tolerance that is related to primary datum feature A, secondary datum feature C, and tertiary datum feature F. The tolerance zone is to be .024" diameter regardless of feature size.

_____

52. Complete a feature control frame for a position tolerance that is related to primary datum feature D, secondary datum feature C, and tertiary datum feature G. The tolerance zone is to be .031″ diameter when the feature is at maximum material condition.

53. Draw the shown tolerance specification in an acceptable location that indicates the tolerance applies to all four holes. Make the necessary dimensions basic.

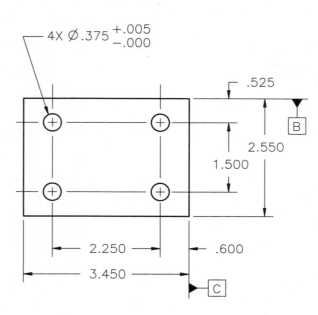

Name _____

54.   A drawing is given and below the drawing are sketches of two produced parts with surface variations exaggerated. Assume the holes are produced exactly on the true positions defined by the drawing. Show dimensions on the produced parts to indicate how the location dimensions are measured on each of the given parts. Show any datum planes that may be needed.

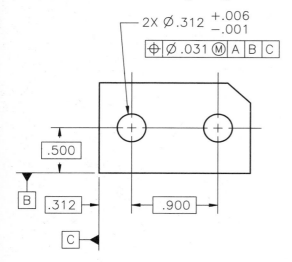

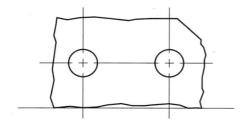

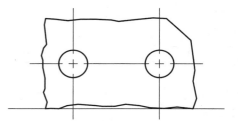

55. Complete calculations to determine the allowable position tolerance for each of the applications shown in the table. Each of the applications is for a floating fastener. Insert your answers in the given table.

SPECIFIED HOLE DIA	FASTENER DIA AT MMC	ALLOWABLE POSITION TOLERANCE AT MMC
.221±.003	.190	
.219±.002	.190	
.282±.004	.250	

56. Complete the given table. All problems are for a floating fastener application.

HOLE DIA AT MMC	FASTENER DIA AT MMC	ALLOWABLE POSITION TOLERANCE AT MMC
.189	.164	
	.190	.031
.279		.029

57. Complete the given table. All problems are for a fixed fastener application.

CLEARANCE HOLE DIA AT MMC	FASTENER DIA AT MMC	ALLOWABLE POSITION TOLERANCE AT MMC
.282	.250	
.218		.014
	.312	.021

Name _____

58. Show your calculations. Enter the X and Y variations for each produced hole in the two tables provided. Calculate the amount of X and Y variation from true position for each hole and enter the variation data in the two tables. Use one or both of the following methods to determine if the produced holes are in acceptable locations.

Solution Method 1. Plot the hole locations on the given grid. Label each hole location with the hole identification number. Draw circles to represent tolerance zone diameters. Note each hole location as acceptable or unacceptable. Each grid space equals .001″.

Solution Method 2. Calculate and enter in the given table the bonus tolerance for each hole. Calculate and enter in the table the allowable position tolerance for each produced hole. Determine by calculation or conversion table the diameter of position variation for each hole and enter the value in the table. Enter in the table whether to accept or reject the hole.

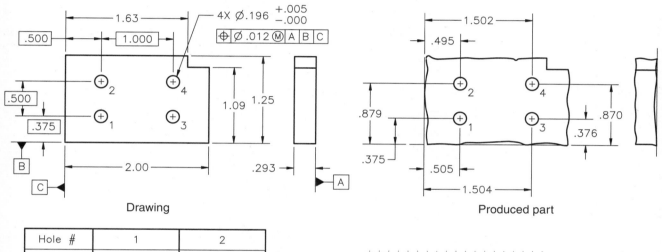

Drawing   Produced part

Hole #	1		2	
Diameter	.199		.201	
	X	Y	X	Y
Measured Location				
Drawing Dimension	.500	.375	.500	.875
Variation				

Hole #	3		4	
Diameter	.200		.200	
	X	Y	X	Y
Measured Location				
Drawing Dimension	1.500	.375	1.500	.875
Variation				

Measured hole data

Plotted coordinate variations and position tolerance zones

Hole #	Bonus Tolerance	Allowable Position Tolerance	Measured Position Variation	Accept or Reject
1	.003	.015	.0100	Accept
2				
3				
4				

59. Complete the detail drawings of the two given parts to the extent required to define hole location requirements. Hole sizes and fastener sizes are provided. Select and identify datums. Dimension the true positions of the holes. Calculate and apply position tolerances that ensure the two parts can be assembled. Use projected tolerance zones if needed. The limits of size for the dowel pin are .1876 to .1878 diameter and may be rounded off to .188 diameter for calculating the position tolerances. The bolt should be assumed to have an MMC size of .250 diameter.

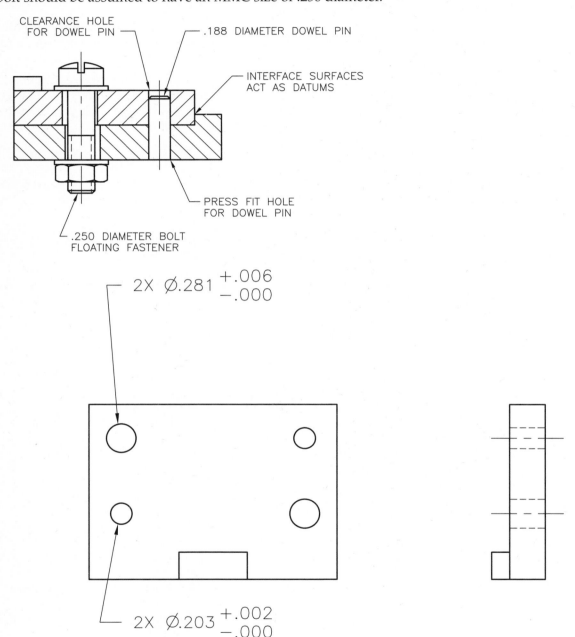

Name _____

2X ⌀.281 $^{+.006}_{-.000}$

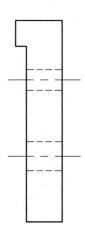

2X ⌀.1865 $^{+.0008}_{-.0000}$

60. Complete the hole specification including a position tolerance of .018" diameter at MMC relative to primary datum feature A, secondary datum feature B, and tertiary datum feature C. Apply the tolerance specification in such a manner that the tolerance is applicable to both the hole and counterbore.

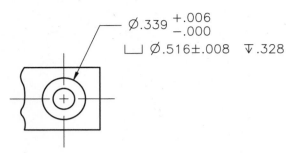

$\emptyset.339\ ^{+.006}_{-.000}$

⌴ $\emptyset.516\pm.008$ ▽ .328

61. Redraw the feature control frame to specify a projected tolerance zone that extends .375".

62. Apply a position tolerance on the given slot to permit .045" location variation in the X axis and a .015" in the Y axis.

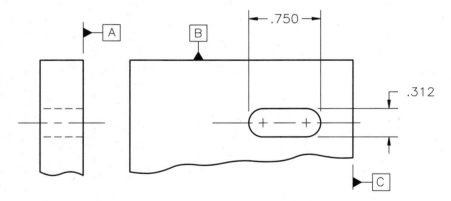

Chapter 9

Position Tolerancing—
Expanded Principles,
Symmetry, and Concentricity

Name _____ Date _____ Class _____

Reading

Read Chapter 9 of the *GD&T: Application and Interpretation* textbook prior to completing the review exercises.

Objectives

A combination of activities is required to achieve the following objectives. Completing the reading assignment and the following review exercises is an important part of achieving the objectives. Familiarization with the objectives prior to completion of the reading assignment and review exercises will make mastery of the objectives easier. After completing the reading assignment and completing the review exercises, you will be able to:

▼ Explain functional gaging methods for checking hole position tolerances specified at MMC.
▼ Specify and explain composite position tolerance specifications.
▼ Explain the effect of using identical datum feature references in multiple position tolerance specifications.
▼ Specify separate pattern requirements for groups of features not acting as a single pattern.
▼ Specify position tolerances for in-line holes.
▼ Specify tolerances to control symmetry.
▼ Control coaxial features with position or concentricity tolerances, depending on the given application.

Review Exercises

Place your answers in the spaces provided. Show all calculations for problems that require mathematical solutions.

Multiple Choice

_____ 1. A single-segment (single-line) position tolerance specification establishes tolerance zones that have _____ relative to the referenced datums.
A. fixed positions
B. no location requirement
C. only a fixed orientation
D. no orientation requirement

_____   2. In a composite position tolerance, the pattern-locating tolerance is
specified _____ the feature-relating tolerance.
A. above
B. below
C. either above or below
D. in a separate feature control frame from

_____   3. The feature-relating tolerance is always _____ than the pattern-locating
tolerance in a composite position tolerance specification.
A. smaller
B. larger
C. equal to or less
D. equal to or greater

_____   4. Referencing primary and secondary datum surfaces in the second
segment (second line) of a composite tolerance specification constrains
rotational degrees of freedom relative to the datums but does not
constrain _____ relative to the datums.
A. part verification
B. angularity
C. translation
D. None of the above.

_____   5. No _____ is created when two position tolerance symbols are shown in a
two segment (two line) feature control frame.
A. valid specification
B. position tolerance specification
C. composite tolerance specification
D. All of the above.

_____   6. The complexity of a functional gage may be impacted by the number
of _____.
A. features being checked
B. tolerance requirements placed on the features
C. referenced datums
D. All of the above.

_____   7. An MMB modifier on a _____ datum feature of size reference always
requires the virtual condition of the datum feature to be used to establish
the datum location.
A. primary or secondary
B. secondary or tertiary
C. primary or tertiary
D. All of the above.

_____   8. The primary characteristic on a drawing that determines whether all
holes belong to one or more patterns is the _____.
A. datum feature references in the position tolerance specifications
B. grouping of holes
C. hole size
D. manner in which hole location dimensions are applied

Name _____

_____   9.   Coaxial (or in-line) holes _____ when using a position tolerance to specify a tolerance that controls the in-line condition.
A.   must have the same diameter
B.   may have different diameters
C.   must have one hole referenced as a datum
D.   None of the above.

_____   10.   _____ tolerances should only be used when it is necessary to control a derived median line RFS relative to a datum axis RMB.
A.   Position
B.   Concentricity
C.   Runout
D.   Composite position

True/False

_____   11.   *True or False?* Parts inspection may be accomplished by using functional gages to check position tolerances that are specified at MMC.

_____   12.   *True or False?* In composite position tolerances, the feature-relating tolerance defines allowable feature-to-feature positions.

_____   13.   *True or False?* The true positions of a feature-relating tolerance zone framework must all be within the pattern-locating tolerance zones.

_____   14.   *True or False?* A feature-relating tolerance zone framework must be properly oriented (rotational degree of freedom constrained) relative to the primary datum that is referenced in the second line of a composite position tolerance specification.

_____   15.   *True or False?* If the first set of measurements for a pattern of holes does not meet the feature-relating tolerance specification when paper gaging, different holes within the pattern may be used to establish a coordinate system for another set of measurements.

_____   16.   *True or False?* Two position tolerance symbols may be used in a two segment feature control frame to specify a composite position tolerance.

_____   17.   *True or False?* A functional gage containing a pin that is sized to the virtual condition of a hole automatically checks the hole location and both size limits for the hole.

_____   18.   *True or False?* All references to datum features of size must include the maximum material boundary modifier when specifying composite position tolerances.

_____   19.   *True or False?* The two gages used to check the pattern-locating tolerance and the feature-relating tolerance for a pattern of holes both have the same diameter of gage pins.

_____ 20. *True or False?* All holes are known to act as a single pattern if the holes are the same diameter.

_____ 21. *True or False?* A composite position tolerance, instead of concentricity, applied to two or more coaxial (in-line) holes must contain at least one datum feature reference for the feature-relating tolerance.

_____ 22. *True or False?* Position tolerances are typically applied to coaxial holes when the main concern is assembly of the parts.

_____ 23. *True or False?* Symmetry tolerances should not be applied to any features other than hole patterns.

_____ 24. *True or False?* Concentricity tolerances may be used to define the allowable surface variations of one cylinder relative to a datum axis.

Fill in the Blank

_____ 25. A _____ segment position tolerance specification made RFS requires all hole locations be within position tolerance zones that are all the same size.

_____ 26. In composite position tolerances, the _____ locating tolerance establishes the hole pattern position requirements relative to the datum references frame.

_____ 27. The _____ segment of a composite position tolerance always specifies the pattern-locating tolerance.

_____ 28. The _____ segment of a composite position tolerance has the same effect as a single segment position tolerance specification.

_____ 29. Paper gaging the feature-relating tolerance for a pattern of holes may be accomplished by using one of the holes as the _____ for measurements within the pattern.

_____ 30. A functional gage for verifying hole locations automatically permits utilization of any allowable bonus tolerance since gage pins are sized to the _____ of the holes being checked.

_____ 31. An MMB modifier on a primary datum feature reference requires the _____ size of the datum feature be used to establish the datum location if Rule #1 is applicable to the feature.

_____ 32. Placing the words _____ under a position tolerance specification results in the associated group of holes acting as a separate pattern from any other holes or features.

_____ 33. If two groups of holes are toleranced with composite position tolerances that reference different datums, _____ patterns of features are created.

_____ 34. Symmetry requirements that apply at MMC are specified using the _____ symbol.

_____ 35. Concentricity is always specified with the tolerance applicable _____ of feature size.

Name _____

Short Answer

36. What requirements apply to the specification of datum features in the second segment of a composite position tolerance?

_____

_____

37. Explain the feature-relating tolerance zone framework requirement for a composite position tolerance specification that is applied to a pattern of holes when no datum reference is shown in the second segment.

_____

_____

_____

_____

38. Why are two holes in a hole pattern used to establish a coordinate system when making measurements to check the feature-relating tolerances?

_____

_____

_____

39. What is a functional gage?

_____

_____

_____

40. What must be accomplished with the datum simulator if the outside diameter of a shaft is referenced as a datum feature with no modifier applied to the reference?

_____

_____

_____

41. When features are dimensioned and toleranced according to the current standard, what indicates that features belong to a single pattern?

_____

_____

_____

42. Why is it possible to dimension a hole pattern without showing a dimension from the pattern of holes to a datum feature when a symmetry position tolerance is specified?

_____

_____

43. What tolerance types are preferable to concentricity for controlling coaxial features?

_____

_____

44. If implied datums are used, what is the risk related to how datums might be assumed in machining and inspection of the part?

_____

_____

_____

Application Problems

All application problems are to be completed using correct dimensioning and tolerancing techniques. Show any required calculations. If your work is sketched, adequate care should be taken to make your answer easily readable.

45. Complete a composite position tolerance specification that creates a pattern-locating tolerance of .036″ diameter at MMC relative to datum features A primary, B secondary, and C tertiary, and a feature-relating tolerance of .011″ diameter at MMC relative to primary datum feature A.

46. Complete the given tolerance specification and identify the two lines of the feature control frame.

⌖ | $\emptyset$ .029 Ⓜ | B | C | E |
   | $\emptyset$ .015 Ⓜ |

Name _____

47. The pattern-locating tolerance zone framework and the pattern-locating tolerances are shown on the illustrated part. Holes are not shown. Show one possible location of the feature-relating tolerance zone framework that does not coincide with the pattern-locating tolerance zone framework. Also show the feature-relating tolerance zones. Show one permissible point for the center location of each hole.

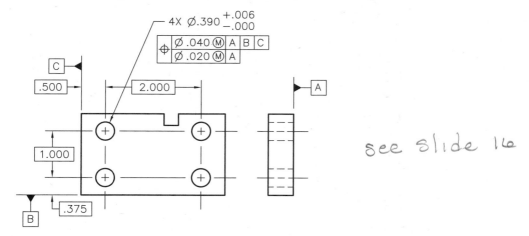

see slide 16

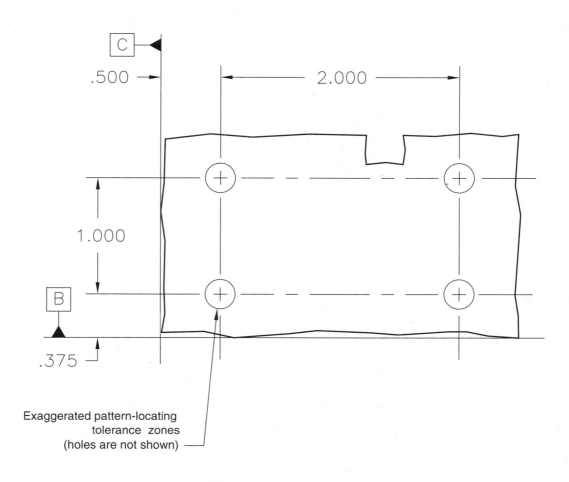

Exaggerated pattern-locating
tolerance zones
(holes are not shown)

48. The pattern-locating tolerance zone framework and the pattern-locating tolerances are shown on the given part. Show one possible location of the feature-relating tolerance zone framework that does not coincide with the pattern-locating tolerance zone framework.

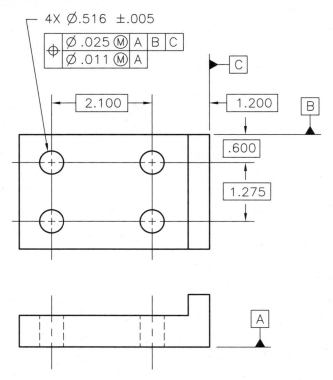

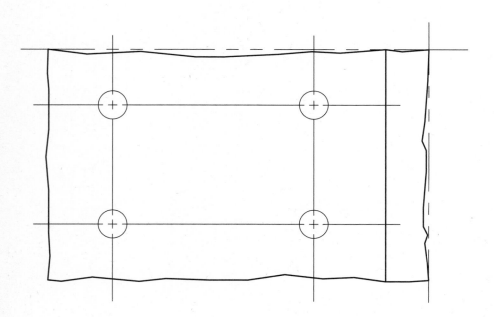

Name _____

49. Complete the steps necessary to prove acceptability or rejection of the given part using paper gaging techniques. Fill in the blanks in the table. Plot the position variations on the provided grid. Label the concentric circles to indicate allowable position tolerance and corresponding hole sizes. Answer the questions in the figure.

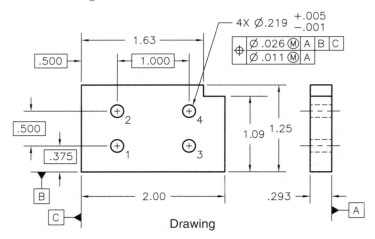

Drawing

HOLE—TO—HOLE LOCATION VARIATION

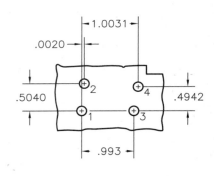

Hole #	1		2		3		4	
Diameter	.222		.223		.221		.223	
	X	Y	X	Y	X	Y	X	Y
Measured Location	0	0	.0020	.5040	.9930	0	1.0031	.4942
Drawing Dimension	0	0	0	.500	1.000	0	1.000	.500
Variation								

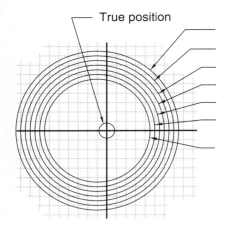

True position

Hole-to-hole relative positions

_____ Is the given location of the concentric circles allowed?

_____ Is bonus tolerance required to make any of the holes acceptable?

_____ Is the feature-relating tolerance met?

50. Design a functional gage that checks the hole positions in the given part. Do not apply gage
tolerances. Superimpose the gage on the given part where the part is shown with phantom lines.

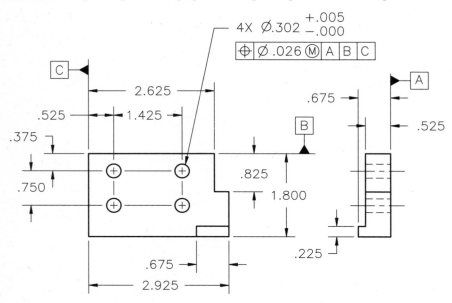

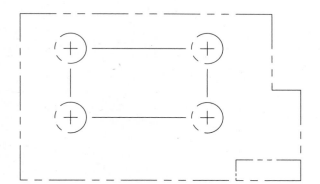

Name _____

51. Calculate the diameter of a pin that establishes the secondary datum for the shown position tolerance specifications.

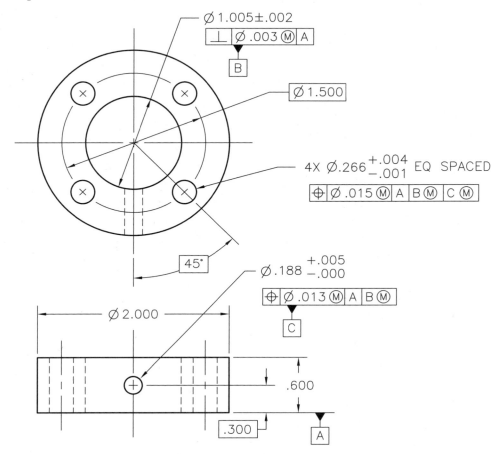

52. Complete a drawing of the gage(s) needed to verify the feature-relating tolerance for all the holes in the given part.

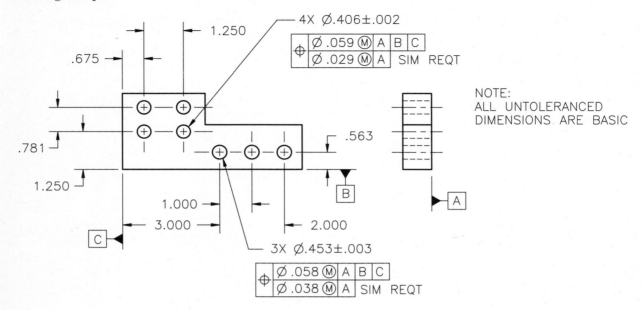

Name _____

53. Complete a drawing of the gage(s) needed to verify the feature-relating tolerance for all the holes in the given part.

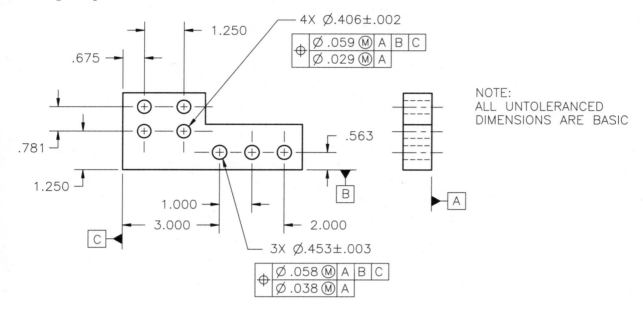

NOTE:
ALL UNTOLERANCED
DIMENSIONS ARE BASIC

54. Apply a composite tolerance to permit a .155" plus or minus .001" diameter shaft to pass through the holes. The shaft must be located within .025" diameter at MMC relative to datum A primary, B secondary, and C tertiary.

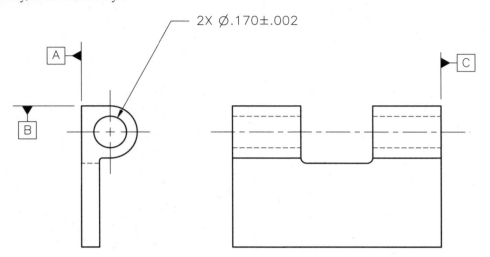

55. Sketch a simple gage that verifies the shown position tolerance.

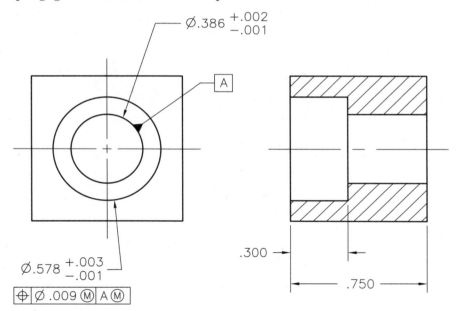

Name _____

56. Apply any additional dimensions and tolerances needed to define hole locations that are symmetrically located to the slot within a .026″ diameter zone when the holes and slot are at MMC. Datum A is primary, the slot secondary, and one end of the part tertiary.

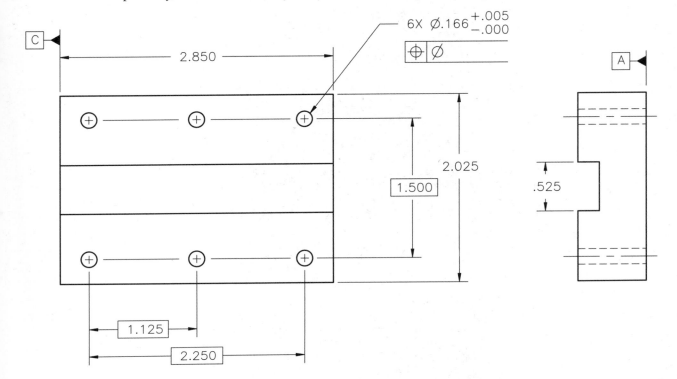

NOTES

Runout

Name _____ **Date** _____ **Class** _____

Reading

Read Chapter 10 of the *GD&T: Application and Interpretation* textbook prior to completing the review exercises.

Objectives

A combination of activities is required to achieve the following objectives. Completing the reading assignment and the following review exercises is an important part of achieving the objectives. Familiarization with the objectives prior to completion of the reading assignment and review exercises will make mastery of the objectives easier. After completing the reading assignment and completing the review exercises, you will be able to:

▼ Describe the two types of runout tolerances.
▼ Complete an interpretation drawing showing how each of the runout tolerances is measured.
▼ Apply both types of runout tolerances on circular features and face surfaces.
▼ Specify runout tolerances using multiple datum feature references.
▼ Limit the area of application for a runout tolerance.

Review Exercises

Place your answers in the spaces provided. Show all calculations for problems that require mathematical solutions.

Multiple Choice

_____ 1. _____ runout includes the variation across an entire surface.
    A. Cylindrical
    B. Total
    C. Face surface
    D. Circular

_____ 2. Circular runout may be measured on any _____ that has circular elements.
    A. cone
    B. cylinder
    C. flat surface
    D. All of the above.

_____   3. A circular runout symbol has _____ arrows.
    A. one
    B. two
    C. either one or two
    D. None of the above.

_____   4. The material condition that always applies to runout tolerances is _____.
    A. MMC
    B. LMC
    C. RFS
    D. All of the above.

_____   5. Runout tolerance specifications must include a _____.
    A. datum feature reference
    B. MMC or LMC modifier
    C. three place decimal tolerance value
    D. None of the above.

_____   6. Datum reference B-C indicates _____.
    A. one datum created by two datum features
    B. two datums created by two datum features
    C. a primary and secondary datum
    D. a single datum created by one datum feature that is identified with the letters B and C

_____   7. A(n) _____ line may be used to indicate a limited area of application for a tolerance specification.
    A. object
    B. center
    C. phantom
    D. chain

True/False

_____   8. *True or False?* Runout may only occur on a cylindrical surface.

_____   9. *True or False?* One runout reading taken at a cross section on a 3.00″ long shaft is adequate to verify a circular runout specification for the 3.00″ shaft.

_____   10. *True or False?* Runout tolerances applied to internal features require notations to explain what the specification means.

_____   11. *True or False?* One datum reference is all that is ever needed for any runout tolerance specification.

_____   12. *True or False?* A runout tolerance may not exceed the size tolerance on the controlled feature.

Name _____

Fill in the Blank

_____ 13. Runout is the amount of _____ variation that is allowed relative to an axis of rotation.

_____ 14. When using a dial indicator for inspection of runout, the part must be _____ on an axis to make the runout measurements.

_____ 15. Two features acting together to establish a single datum axis, such as A–B, through those features is referred to as _____ datum features.

_____ 16. Runout tolerances applied to the outside diameter of a gear blank are measured by rotating the workpiece on the datum axis with a _____ against the outside diameter of the gear blank.

_____ 17. When both a primary and secondary datum reference are shown in a runout tolerance specification, usually the datum features include one _____ surface and one face (flat) surface.

_____ 18. _____ runout is the variation across an entire surface relative to an axis of rotation.

Short Answer

19. Explain how a circular runout requirement is checked on a cylindrical feature when using a dial indicator.

_____

_____

_____

20. Why is a diameter symbol not used in runout tolerance specifications?

_____

_____

21. What is achieved by the application of a total runout tolerance on a surface that is perpendicular to the axis of rotation?

_____

_____

_____

22. Give one reason why there might be a datum reference such as D–E in a runout tolerance.

_____

_____

_____

_____

23. How may a face surface, as a secondary datum reference, be beneficial when a runout tolerance is referenced to a primary datum axis?

_____

_____

_____

24. List two geometric shapes other than a cylinder or flat surface that may be controlled with circular runout.

_____

_____

Application Problems

All application problems are to be completed using correct dimensioning techniques. Show any required calculations.

25. Show two ways to apply a circular runout tolerance specification of .006″ on the small diameter relative to datum axis A.

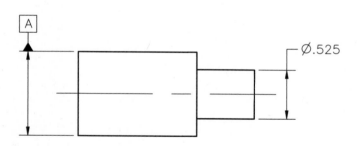

Name _____

26. Sketch a setup and measurement method that may be used to check the runout tolerance. Also show the acceptable tolerance zone at multiple locations on the feature.

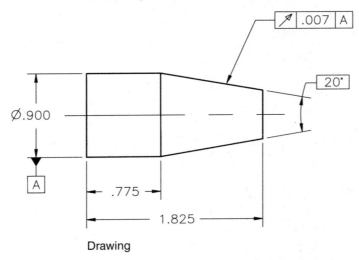

Drawing

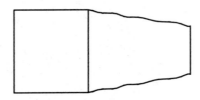

27. Complete a feature control frame that specifies a circular runout tolerance of .008″ relative to an axis established by datum feature C.

28. Sketch a setup and measurement method that may be used to check the runout tolerance. Also show the acceptable tolerance zone at multiple locations on the feature.

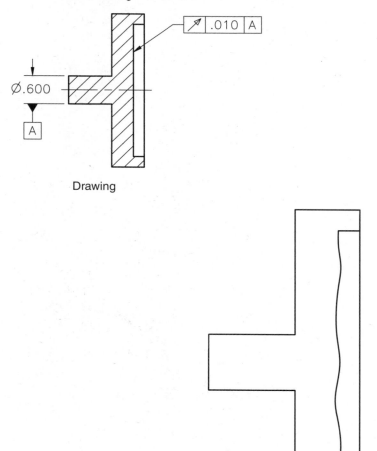

Drawing

29. Apply the necessary symbology to control the circular runout of the .375″ diameter to a value of .006″ relative to an axis established by the two .250″ diameter bearing surfaces.

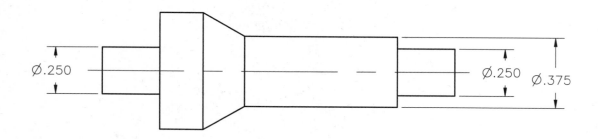

Name _____

30. Sketch a setup and measurement method that may be used to check the runout tolerances. Also show the acceptable tolerance measurements.

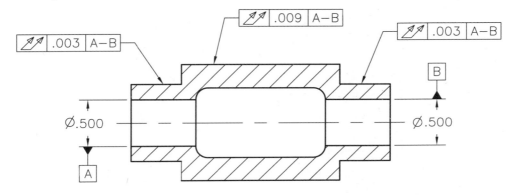

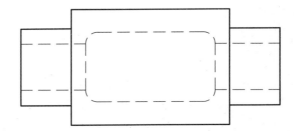

31. Sketch a setup and measurement method that may be used to check the runout tolerance. Also show the acceptable tolerance measurements.

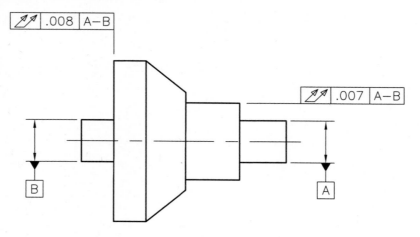

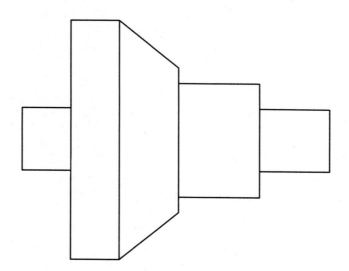

Chapter 11

Profile

Name _____ **Date** _____ **Class** _____

Reading

Read Chapter 11 of the *GD&T: Application and Interpretation* textbook prior to completing the review exercises.

Objectives

A combination of activities is required to achieve the following objectives. Completing the reading assignment and the following review exercises is an important part of achieving the objectives. Familiarization with the objectives prior to completion of the reading assignment and completing the review exercises after reading will make mastery of the objectives easier. After completing the reading assignment and completing the review exercises, you will be able to:

▼ Define line and surface profile tolerances.
▼ Apply profile tolerances to define allowable variation within a limited zone on a feature or for all of a feature.
▼ Apply profile tolerances to extend all around the profile shown in a drawing view.
▼ Complete profile tolerance specifications to achieve any of the possible levels of control.
▼ Sketch the tolerance zone created by profile tolerance specifications.
▼ Specify coplanarity requirements using profile tolerances.
▼ Identify profile tolerances as the means for specifying allowable variation for conical surface form, orientation, and location.
▼ Draw a composite profile tolerance specification.

Review Exercises

Place your answers in the spaces provided. Show all calculations for problems that require mathematical solutions.

Multiple Choice

_____ 1. Only the _____ is different between the format of a line profile and a surface profile feature control frame.
   A. datum referencing method
   B. use of basic dimensions
   C. all around symbol usage
   D. tolerance symbol

_____   2. If a profile tolerance _____, the profile tolerance does not control the
location or orientation of the toleranced surface.
   A.   is a line profile control
   B.   is a surface profile control
   C.   does not include datum feature references
   D.   All of the above.

_____   3. Profile of a line is similar to _____ tolerances since individual line elements
are controlled separately.
   A.   straightness
   B.   flatness
   C.   perpendicularity
   D.   angularity

_____   4. A profile tolerance may be applied to less than a whole surface by defining
and referencing _____.
   A.   limits of size
   B.   limits of application
   C.   dual requirements
   D.   datums

_____   5. Unless indicated otherwise, profile tolerances are assumed to be _____.
   A.   unilateral
   B.   equally disposed bilateral
   C.   all around
   D.   unequally disposed

_____   6. Unidirectional (unequally disposed) profile tolerances may be applied to
control _____.
   A.   form
   B.   form and orientation
   C.   form, orientation, and size
   D.   All of the above.

_____   7. Datum feature references are included in a profile tolerance where _____
is to be controlled.
   A.   form
   B.   size and form
   C.   form, orientation, and location
   D.   None of the above.

_____   8. A basic dimension is used to locate a feature controlled by a profile
tolerance where _____ is to be controlled.
   A.   form
   B.   form and size
   C.   position
   D.   Either B or C.

_____   9. To control form only, _____ datum feature reference(s) must be used.
   A.   no
   B.   one
   C.   two
   D.   three

Name _____

_____ 10. A profile tolerance may be specified not to extend across an entire feature by indicating a _____.
A. limited extent of application
B. drawing the tolerance zone
C. drawing a line to one side of the basic profile
D. None of the above.

_____ 11. A surface profile tolerance applied to a cone should include _____ to establish location, orientation, form, and size requirements.
A. no datum feature references
B. no basic location dimensions
C. neither datum feature references or basic location dimensions
D. datum feature references and basic location dimensions

True/False

_____ 12. *True or False?* Profile tolerances are always specified with the MMC modifier.

_____ 13. *True or False?* A curved surface must be defined by basic dimensions when a profile tolerance is applied to the surface.

_____ 14. *True or False?* Surface profile may only be used to control the form of a curved surface.

_____ 15. *True or False?* Even when an all around symbol is used, profile tolerances do not extend past abrupt changes in direction.

_____ 16. *True or False?* There is no number shown following the unequally disposed symbol if the entire profile tolerance zone goes inside the material relative to the basic profile of the surface.

_____ 17. *True or False?* When used, unequally disposed profile tolerances must be applied to permit a plus size tolerance rather than a minus size tolerance.

_____ 18. *True or False?* A feature controlled by a profile tolerance may be located by a basic dimension and when basic location dimensions are shown the profile tolerance must include appropriate datum feature references.

_____ 19. *True or False?* A composite profile tolerance may be used to specify a small tolerance for form of a surface and a large tolerance for the form, orientation, and location relative to one or more datums.

_____ 20. *True or False?* One method of specifying coplanarity of multiple flat surfaces is to apply a flatness tolerance.

Fill in the Blank

_____ 21. There are at least _____ segments in a composite profile tolerance.

_____ 22. _____ profile tolerance may be applied to a surface, but it only controls individual line elements on the surface.

_____ 23. Profile tolerances apply along the entire surface to which they are applied, and typically the limits of the surface are defined by _____ changes in direction.

_____ 24. In past practices and today as an alternate practice a(n) _____ line is drawn to one side of a feature outline to indicate that a profile tolerance is unilateral.

_____ 25. Whether or not _____ are shown in a feature control frame establishes whether the profile tolerance controls only form or if it controls form, orientation, and location.

_____ 26. No _____ is shown in a profile tolerance specification when controlling form only.

_____ 27. Dimensions that define the shape of a surface must be _____ if a profile tolerance is applied.

Short Answer

28. Profile tolerances are typically attached to a controlled surface in what manner?

_____

_____

29. How would a profile tolerance that applies all the way around a feature profile be indicated?

_____

_____

_____

30. Describe an unequally disposed profile tolerance and how it is applied on a drawing.

_____

_____

_____

_____

_____

Name _____

31. Explain the impact of applying a basic dimension for the location of a surface that has a profile tolerance including datum feature references in the feature control frame.

_____

_____

_____

_____

32. Place an *X* by each characteristic that affects the required level of control on a feature.

    _____ Line or surface profile symbol

    _____ Datum feature references

    _____ Total area of the controlled surface

    _____ Basic location dimensions

    _____ Curved or flat surface

33. How can a coplanarity requirement for multiple flat surfaces be specified?

_____

_____

Application Problems

All application problems are to be completed using correct dimensioning techniques. Show any required calculations.

34. Apply a line profile tolerance that only controls the form of the curved surface within a boundary .025″ wide.

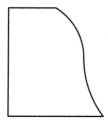

35. Show the tolerance zone created by each of the given tolerance specifications. Superimpose the tolerance zone on the given drawing.

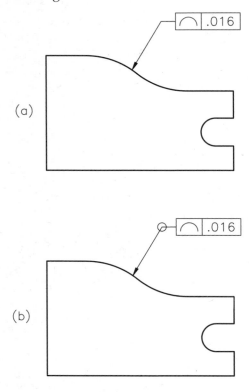

(a)

(b)

(c)

36. Complete the drawing to the extent necessary to control the surface profile all around the perimeter of the part within a boundary .040″ wide. Indicate basic dimensions where they are needed.

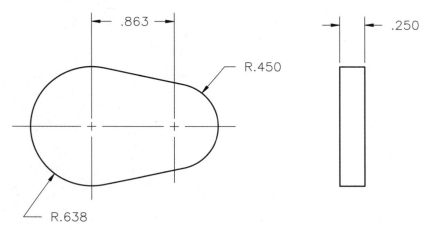

Name _____

37. Control the surface profile of the given slot all around within an unequally disposed zone .015" to the inside of the material (reduces the material). Also, control both location and orientation of the profile tolerance zone to three datums. Indicate basic dimensions where they are needed.

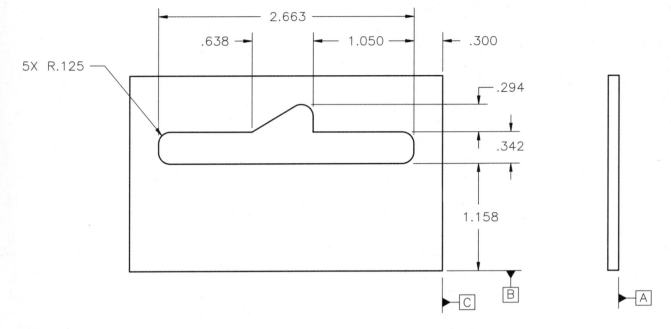

38. Show the tolerance zone for the given slot. Superimpose the tolerance zone on the given drawing. Dimension the width of the slot and the offset from the true profile.

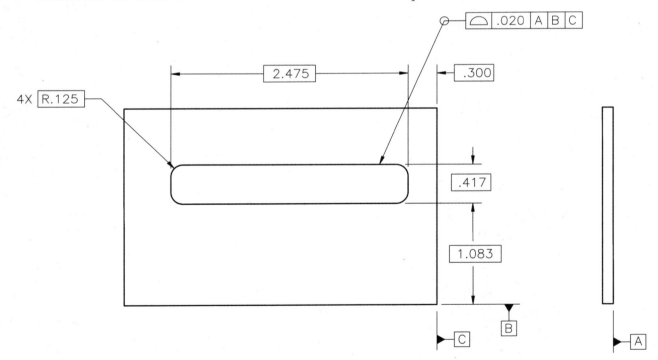

39. Control the form and size of the punched hole within a surface profile of .010″. Use position at MMC to specify a location tolerance of .020″. Identify and use datum features as needed. Indicate basic dimensions as needed.

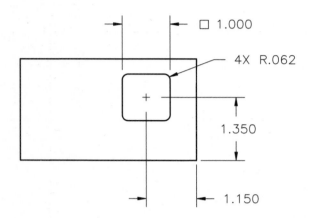

40. Require flat and coplanar bosses within an .008″ tolerance zone. Allow location and parallelism within ±.015″ relative to the bottom surface. Use composite profile to specify the requirements.

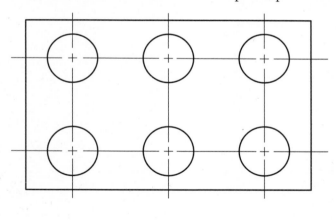

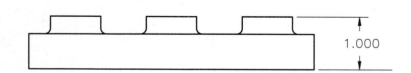

Name _____

41. Require flat and coplanar bosses that are located within an .008″ tolerance zone. Require the zone to be centered 1.000″ from datum A and parallel to datum A.

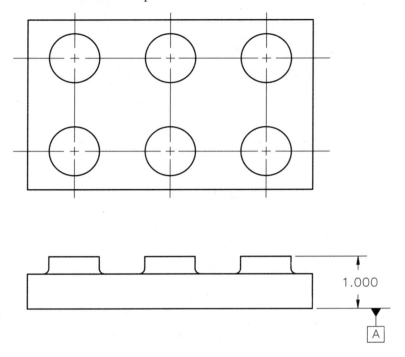

42. Specify a tolerance zone that controls the cone surface size, form, orientation, and location relative to datum axis A and datum plane B within a boundary that is .018″.

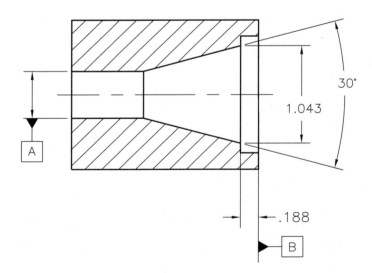

43. Apply a composite profile tolerance to control coplanarity to within .005″ and location and parallelism to datum A within .025″. Apply basic dimensions where needed.

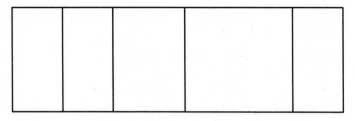

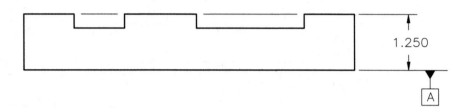

1.250

A

Chapter 12

Practical Applications and Calculation Methods

Name _____ Date _____ Class _____

Reading

Read Chapter 12 of the *GD&T: Application and Interpretation* textbook prior to completing the review exercises.

Objectives

A combination of activities is required to achieve the following objectives. Completing the reading assignment and the following review exercises is an important part of achieving the objectives. Familiarization with the objectives prior to completion of the reading assignment and review exercises will make mastery of the objectives easier. After completing the reading assignment and completing the review exercises, you will be able to:
▼ Calculate position tolerances when more than two parts are stacked in a floating fastener or fixed fastener application.
▼ Distribute the total available position tolerance between features to which position tolerances are applied.
▼ Specify projected tolerance zones for fixed feature locations to prevent interference conditions.
▼ Determine the amount of tolerance accumulation in a simple assembly.
▼ Properly use zero position tolerances at MMC to increase manufacturing freedom.
▼ Apply paper gaging techniques to determine if a produced part meets drawing requirements.
▼ Explain allowable position tolerance effects resulting from datum references at MMB.

Review Exercises

Place your answers in the spaces provided. Show all calculations for problems that require mathematical solutions.

Multiple Choice

_____ 1. If edges of stacked parts in a floating fastener condition must align, then the edges are referenced as _____ in the tolerance specification.
  A. origins
  B. datum features
  C. primary surfaces
  D. mated surfaces

_____ 2. When using the formula T = H − F to calculate one position tolerance value for both parts in a floating fastener condition, the holes _____.
    A. must be the same specified size
    B. may be different specified sizes
    C. must be smaller than the value used for H
    D. None of the above.

_____ 3. To increase the allowable amount of tolerance, what can be specified when alignment of datum features is not required?
    A. Specify a composite position tolerance.
    B. Specify a bonus tolerance.
    C. Specify a large pattern-locating tolerance.
    D. Both A and C.

_____ 4. In a floating fastener application, the correct amount of position tolerance for a .190″ diameter bolt and .228″ MMC diameter hole is _____ inch.
    A. .014
    B. .019
    C. .028
    D. .038

_____ 5. Two of three stacked parts must have _____ in a fixed fastener condition.
    A. threads
    B. press fit sizes
    C. clearance holes
    D. None of the above.

_____ 6. The allowable position tolerance that may be applied to each part in a fixed fastener application is _____ inch if the clearance hole is .282″ diameter MMC and a .250″ diameter bolt is used.
    A. .014
    B. .016
    C. .028
    D. .032

_____ 7. Generally, a threaded hole is given _____ the clearance hole to improve producibility.
    A. more position tolerance than
    B. the same position tolerance as
    C. less position tolerance than
    D. None of the above.

_____ 8. A projected tolerance zone is indicated by a(n) _____.
    A. letter P inside a circle
    B. arrow pointing to the outside of the part
    C. note under the feature control frame
    D. All of the above.

_____ 9. A projected tolerance zone is typically specified to extend a distance equal to the _____.
    A. fastener length
    B. length of the fixed segment of the fixed fastener
    C. clearance feature length
    D. fastener diameter

Name _____

_____ 10. A hole size specification of .210″ minimum and .216″ maximum diameter
has a position tolerance specification of .020″ diameter MMC. A .190″
diameter floating fastener passes through the hole. If the hole is produced
at .208″ diameter and has a position variation of .012″ diameter, what
should be done?
A. Accept the part since it meets the specification.
B. Accept the part since it is functional.
C. Reject the part and throw it away.
D. Rework the part to make the hole an acceptable diameter.

_____ 11. For a floating fastener application, a hole size specification of .385″
minimum and .395″ maximum diameter has a position tolerance
specification of .010″ diameter MMC. If the position tolerance is changed
to .000″ diameter MMC, a minimum hole diameter of _____ inch must
be specified with the maximum size limit remaining .395″.
A. .375
B. .380
C. .385
D. .390

_____ 12. Concentric circles used to paper gage a feature-relating tolerance
requirement _____ relative to the graph origin.
A. must be centered
B. are free to float
C. are offset a distance equal to the location of the nearest hole
D. None of the above.

_____ 13. If a single segment tolerance specification is applied to a single flat
surface and does not include any datum references, the tolerance is
either _____.
A. form or runout
B. form or orientation
C. form or profile
D. profile or orientation

_____ 14. A single feature may require a maximum of _____ level(s) of control, each
specified in a separate feature control frame.
A. no
B. one
C. two
D. None of the above.

_____ 15. A flat surface may have a perpendicularity tolerance of .017″ applied to
it and also have a _____ tolerance of .008″ applied to further refine the
surface form.
A. flatness
B. parallelism
C. position
D. circularity

True/False

_____ 16. *True or False?* A floating fastener condition exists only when a maximum of two stacked parts have clearance holes through which a fastener passes.

_____ 17. *True or False?* If the clearance holes in mating parts are the same size, different position tolerance values may be applied on each hole.

_____ 18. *True or False?* If one part is purchased with hole position tolerances already specified by the manufacturer, it is not possible to calculate position tolerances for the mating parts.

_____ 19. *True or False?* A projected tolerance zone extends the full length of the controlled feature plus a projected distance outside the feature.

_____ 20. *True or False?* A specified zero position tolerance at MMC is an error since perfect position is seldom, if ever, achieved.

_____ 21. *True or False?* Even if a part is functionally adequate, the part must be rejected, reworked, or accepted by special procedures if it does not meet drawing requirements.

_____ 22. *True or False?* Paper gaging should only be used for position tolerances specified with the MMC modifier.

_____ 23. *True or False?* Zero position tolerances should not be specified without a material condition modifier.

_____ 24. *True or False?* Paper gaging of the feature-relating tolerance in a composite position tolerance specification may be completed by plotting the hole-to-hole measurements without concern for the hole locations relative to any datums.

Fill in the Blank

_____ 25. Show the formula used to calculate floating fastener condition dimensions for two parts that must have aligned surfaces.

_____ 26. Complete the formula used to calculate unevenly distributed tolerances when both holes are the same specified size. $T_1 + T_2 =$ _____ $- 2F$.

_____ 27. What is the formula for calculating distributed tolerances in a floating fastener application in which two hole sizes are specified?

_____ 28. When three or more parts are stacked in a fixed fastener condition, tolerances are calculated considering _____ parts at a time if the clearance holes are different diameters.

_____ 29. Evenly distributed position tolerances for a fixed fastener condition are calculated using what formula?

Name _____

_____ 30. The total available position tolerance for a fixed fastener condition may be distributed between two parts using what formula?

_____ 31. A _____ tolerance zone controls the location outside of the feature.

_____ 32. A correctly specified zero position tolerance at _____ results in all functionally good parts being acceptable.

_____ 33. A specified hole diameter of .163″ minimum and .168″ maximum has a specified position tolerance of .025″ diameter at LMC. A produced hole of .165″ diameter has an allowable position tolerance of _____ inch diameter.

Short Answer

34. If three stacked parts all have the same diameter clearance holes, how are position tolerances for the holes calculated?

_____

35. If the total allowable position tolerance for a fixed fastener application is .022″, what would be wrong with applying .020″ diameter tolerance on one part and .002″ diameter tolerance on the other part?

_____

_____

_____

_____

36. Describe a fixed fastener condition.

_____

_____

_____

37. Why is the manufacturing process considered when distributing tolerances between two parts in a fixed fastener condition?

_____

_____

_____

38. Why is it sometimes necessary to show the direction that a projected tolerance zone extends?

_____

_____

_____

_____

39. A hole for a .250″ diameter bolt is specified to have a .260″ minimum and .268″ maximum diameter with a position tolerance of .010″ diameter at MMC. What may be done to the hole size and tolerance specifications to maximize manufacturing freedom?

_____

_____

_____

Application Problems

All application problems are to be completed using correct dimensioning techniques. Show any required calculations.

40. Complete a composite position tolerance that may be applied to the pattern of holes in each part. Bolts measuring .250″ diameter pass through the holes. The datum features on the part may be misaligned by .030″.

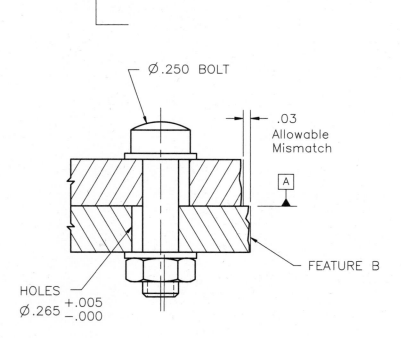

Ø.250 BOLT

.03
Allowable
Mismatch

A

FEATURE B

HOLES
Ø.265 +.005 −.000

Name _____

41. Apply the maximum allowable position tolerance specification on the untoleranced hole. A .375″ diameter bolt passes through the holes.

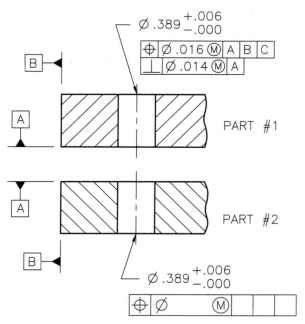

42. An assembly drawing is given. Draw one view of each part that shows the hole patterns. Dimension the hole pattern and apply tolerances for a fixed fastener condition with a .250″ diameter bolt and clearance holes .292″ at MMC.

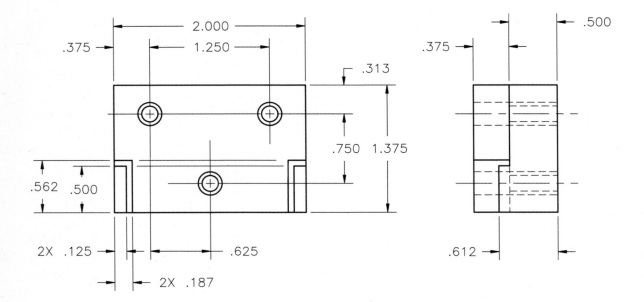

Name _____

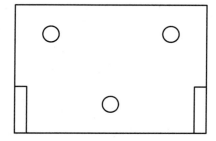 

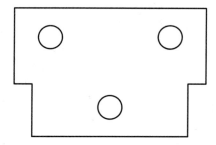 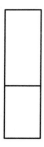

43. Calculate and apply position tolerances for the two given parts. Apply 66% of the total tolerance on the threaded holes.

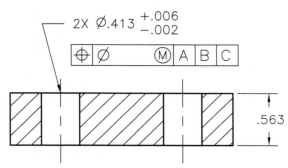

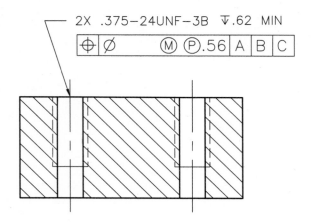

44. Show the tolerance zones for the given holes.

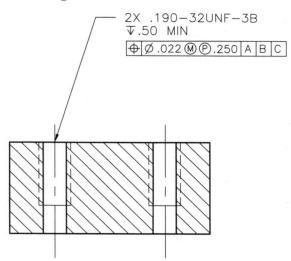

Name _____

45. What is the correct projected distance for a position tolerance applied to the threaded hole? Why is that distance the correct one?

_____

_____

_____

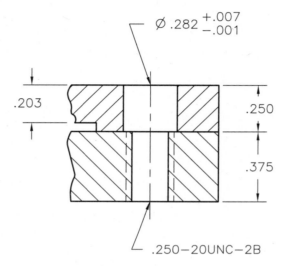

46. Apply a size tolerance of ±.030″ for the given dimension. Require the top surface to be parallel to datum A within .020″ and flat within .009″.

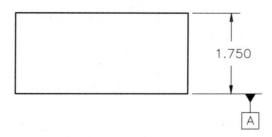

47. Calculate the allowable specified position tolerance for the specified clearance hole on the shown plate. Assume datums are selected to minimize tolerance stackup.

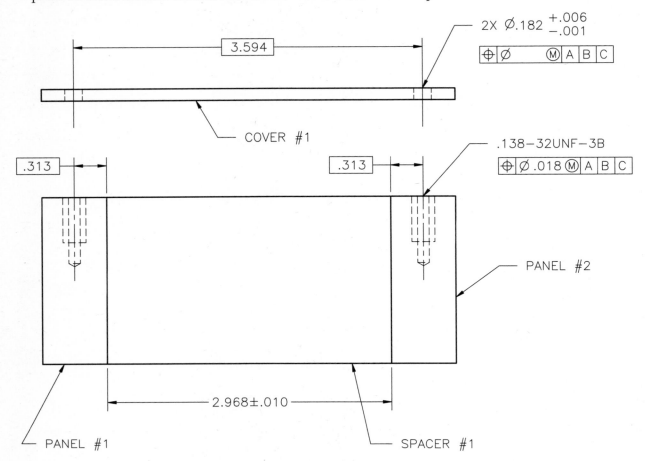